CW00552993

THE MATTER OF AIR

The Matter of Air

Science and the Art of the Ethereal

Steven Connor

REAKTION BOOKS

Published by
Reaktion Books Ltd
33 Great Sutton Street
London EC1V ODX, UK

www.reaktionbooks.co.uk

Printed and bound in Great Britain
by CPI/Antony Rowe, Chippenham, Wiltshire

British Library Cataloguing in Publication Data
Connor, Steven, 1955–
The matter of air : science and art of the ethereal.
1. Air.
2. Air quality.
3. Air –Pollution.
I. Title
551.5'1-DC22

ISBN: 978 1 86189 766 4

Contents

PART TWO:

Atmospherics

six: Haze

seven: Atmospherics

eight: A Grave in the Air

PART THREE:

Absolute Levity

nine: Air's Exaltation

ten: The Fizziness Business

Opening

'Aer', an anonymous Low Countries engraving,
c. 1580s, from a series of *The Four Elements*.

one

Taking to the Air

Implication

It seems we have taken to the air. And in at least three different ways.

First, we have taken to the air as an arena of enquiry. During the last four centuries or so, the air has been intensely and systematically subject to observation, experiment and speculation. It is not as though human beings had been inattentive to the air before that time, for the air had pervaded all aspects of human life, ritual, mythology, medicine, agriculture and technology. As we will see in a moment, the intense activity with regard to the air since the late sixteenth century has sometimes encouraged some ill-advisedly absolute judgements from those concerned with the history of human relations with the air. It has sometimes been supposed that, before the sixteenth century, the air was experienced only as immersion rather than as practical involvement, intuition rather than understanding, or inspiration rather than application, suggesting that the air was in some sense the unconscious itself. Nevertheless, during the period of what we call scientific modernity, we have taken to the air in our tendency to take the air as an object of scientific, technical and philosophical concern.

Second, we have taken to the air in the sense in which a walker takes to the road, a duck takes to water or any creature to its native element. Our understanding of the air has allowed us to occupy it, in ways that had previously been possible only in dream (though to be sure, the air was thickly enough inhabited in dream). But now the air is literally occupied: in air travel, by many different conveyances, from balloons to jet planes, in the exploration of space, which requires the traversal of and re-entry into the atmosphere, in radio, TV and other electromagnetic signals carried through the air, and in the use of X-rays and other forms of radioactivity. Our occupation of the air has even extended temporally backwards and forwards, through the investigation of climate patterns and measurements that allow us to make sensitive forecasts of the weather, and even to some small degree to control it. The fact that we can abstract and even manufacture our own supplies of air and can maintain artificial climates enables us to live in portions of the earth where the enmity of the air would otherwise prevent human habitation, and even to travel out of the air altogether, into the oceans and beyond the atmosphere. 'Can man live elsewhere than in air?' asks Luce Irigaray.[1] One sees what she means, but the answer must be, decidedly, *yes*. The thin air has become unignorably thick, for us, because it is thick with us. Where, for the tens of thousands of years that there have been humans, these creatures have been humble, *humilis*, of the earth – earthy, scuttling around on the ground floor of an unsuspected ocean of air, with occasional feeble aspirations to shin a tree or scale a peak – in the last three or four centuries, we have made a habitation of the air. If we are still yet some way short of being post-human, we have certainly for some time been posthumous.

This leads to the third sense in which we may be said to have taken to the air. For occupying the air as we so much do, and occupied with the air as we so much are, we have begun to find the air more and more expressive of our condition. In seeking to understand our physical composition and that of our world, we have turned to air. We know our bodies, minds and social forms through its light mirror. Human beings have always believed themselves to be in part airy, and have often wanted to believe that their most essential part – their spirits, as they have liked to call them – were aeriform. And yet, in their clinging humility, human beings have viewed the idea of a literal translation into an airy condition with suspicion. For airiness also signifies delusion, insignificance, even madness. When Marx and Engels declared that, under capitalism, 'all that is solid melts into air', they were being prescient, but not approving. Others, more recently, have been inclined to accept this dissolution not as a degradation, but as a fundamental evolution in the system of human relations, though one that would require us to think anew about what we meant by foundations and fundaments. Michel Serres notes that

> [t]he system's 'matter' has changed 'phase', at least since Bergson. It's more liquid than solid, more airlike than liquid, more informational than material. The global is fleeing towards the fragile, the weightless, the living, the breathing.[2]

Serres has formulated this insight in a number of different ways over the last ten or fifteen years. In a remarkable chapter from his 1994 book *Atlas*, entitled 'Temps du monde' (signifying both 'The Time of the World' and 'The Weather of the World'), Serres proposes that we make out from the turbulence of meteorology a metaphor, more, a veritable *mappamundi*, for the movements of information. We inhabit a kind of informational weather, in which the atlas is the territory, or the air-itory: the physics of solids, with their clearly defined edges and outlines, seem scarcely appropriate even to the era of thermodynamics, let alone that of the movement of information. But the physics of 'gaseous, aerial, or viscous states of matter' offer an image that can make 'concrete' the forms that information takes:

> The volatile, mixed elements form material supports for an information which is yet more volatile, and the mingling or modelling of which concurs, better still, with the formation of the Universe, which all this concreteness allows to grow. The logical message forms part of the flow of matter and is born of it: rise up, welcome storms . . . Aphrodite, naked and beautiful, appears from the waves, the Word emerges from the flesh of the world and, in return, creates it as World.[3]

Peter Sloterdijk has echoed some of these arguments in the third volume of his philosophical trilogy *Sphären*, entitled *Schäume* (*Foams*). Discussing the work in an interview, he declared that

> all previous natural languages, including theoretical discourse, were developed for a world of weight and solid substances. They

> are thus incapable of expressing the experiences of a world of lightness and relations . . . Because of their Marxist heritage, critical theorists succumb to the realistic temptation of interpreting the light as appearance and the heavy as essence. Therefore they practice criticism in the old style in that they 'expose' the lightness of appearance in the name of the heaviness of the real. In reality, I think that it is through the occurrence of abundance in the modern that the heavy has turned into appearance – and the 'essential' now dwells in lightness, in the air, in the atmosphere.[4]

Sloterdijk's trilogy reads human history – philosophical, religious, artistic, political – as the successive elaboration of three different kinds of spheres, or spaces of introversion. The first volume concentrates on 'microspherology', the construction of the intimate and elementary spheres, typified by the simplest dyadic relations between the child and the womb, or mother. The second follows through the macrospherological evolution of larger, more inclusive, and metaphysical spheres, typified in imperial conceptions of the One World, or in Marshall McLuhan's notion of the global village. The third volume, *Schäume*, proposes by contrast that the modern world must be understood non-holistically, and in terms of a polyspherology, which will take account, in a 'multifocal, multiperspectival and heterarchical' manner, of the complex aggregations of different spheres that make up the world.[5] Where the governing metaphor for microspherology is the bubble, and the governing metaphor of macrospherology is the globe, the aptest and most versatile metaphor for the polyspherological condition of the modern world is that of foam: 'In place of the philosophical super-soapbubble, of the All-Monad of the unitary world . . . there is a polycosmic agglomeration. This may be described as an assemblage of assemblages, a semi-opaque foam of world-making constructions of space' (*Schäume*, 63–4).

Such a system is characterized by lightness, fragility and ephemerality (foams are never permanent and always in a state of evolution and devolution). Sloterdijk follows through the mixed traditions relating to foam. On the one hand, the foamy, defined as 'air in an unexpected place', in a parody of Mary Douglas's definition of dirt as 'matter out of place' (*Schäume*, 28), signifies the aberrant, the anomalous, even the diabolical. Dreams, fantasies and delusions are frothy, in that they are 'almost nothing and yet not nothing' ('Fast nichts, und doch nicht nichts' – *Schäume*, 27). But Sloterdijk acknowledges an alternative tradition in which the foamy is seen as generative, for example in Hesiod's account of the birth of Aphrodite from the waves, which proposes an etymology of her name as 'foam-born', from the Greek *aphros*, meaning foam (*Schäume*, 42). This alternative tradition, which is evidenced in the prominence of oceanic churning in many cosmogonies, and stays alive through Aristotle's account of the spermatic foam of semen and in the central value accorded to fermentation in alchemy, would suggest the contrary belief that, far from being diabolical, all creation owes its being to the fecundating powers of foam, and that human beings may indeed be '*such stuff as* [wrongly given as 'the'] *foams are made on*' (*Schäume*, 41). The joke is a bit better in German, in which *Schäume* is closer to *Träume*, a coincidence that also prompts Sloterdijk, never one to let slip a paronomasic opportunity, to a parody of Freud's *Traumdeutung*, or interpretation of dreams, in a section of his text he entitles 'Schaumdeutung' (*Schäume*, 32).

Sloterdijk proposes that the modern world requires an aphrology, his name for 'the theory of co-fragile systems' (*Schäume*, 38). Indeed, his emphasis throughout *Schäume* is on the new responsiveness to this polyspheric fragility which characterizes philosophical, artistic and industrial modernity. Fragility, he writes, 'must be thought as the location and mode of that which is most real' (*Schäume*, 9). Since the modern world is able to take seriously the insignificant, the trivial (dreams, for example), modernisation can then be seen

> as a new global process for admitting the fortuitous, the momentary, the vague, the ephemeral and the atmospheric . . . one must speak of a discovery of the indeterminate as a result of which – perhaps for the first time in the history of thought – the not-nothing (*Nicht Nichts*), the almost-nothing, and the formless (*informe*), have gained access to the domain of theorisable realities (*Schäume*, 34, 35)

Now we are airborne, we are, in almost truth, becoming air-born.

Hermetic

This book will concern itself with this process whereby, as a result of explicating the air, we come more and more to merge with it. It will take as its beginning the process of experimental investigation into the mechanics of the air that begins with the work of Galileo and Torricelli, and is extended through the work of Boyle, Gay-Lussac and others, along with the pneumatic chemistry that perhaps may be said to begin with Van Helmont's coining of the word 'gas' in the early seventeenth century and to extend through the major discoveries of Black, Cavendish, Priestley and Lavoisier of the later eighteenth century. This will not, however, be a history of scientific thought. Rather, my concern will be with the ways in which new understandings of the air entered social experience and altered human experiences of their ways of inhabiting the world. I will be interested, not in isolating the air as a specific subject of concern, but with following through some of the ways in which new apprehensions of the air entered into composition with forms of social life and imagining. Ironically, the many different ways in which the air had to be isolated in order to be investigated will be one of the most important vehicles of this composition. I will be interested in the ways in which the scientific understanding of the air has both given rise to new kinds of object and made of the air itself a new kind of object, almost, one might say, a new way of being an object, a light way, a volatile, a virtual, a versatile way.

There are two phases in the experimental investigation of the air. The first is focused on the experiments in what we would now call the physics of the air, which begin in the work of Torricelli, Galileo and others, from the 1640s onwards. This work began to observe and define the air in terms of weights and measures, actions, movements and forces. The principal scientific outcomes of this work are the 'gas laws', governing the relations between pressure and volume (Boyle's law, 1662), between temperature and volume (Charles's law, 1787), and between pressure and temperature (Gay-Lussac's law, 1809). These were all refined and unified by Avogadro's law (1811), which states that equal volumes of gases, at the same temperature and pressure, contain the same number of molecules.

But a purely scientific history of this process, no matter how detailed and specific, may miss out the particular kinds of pressure that were exerted on and through this first century of experiment. Indeed, pressure itself is at the centre of this work. The principal fact of interest to researchers of the sixteenth and seventeenth centuries was that the air had weight and elasticity. The experiments of Boyle and others centred on actions either of evacuation, as in the many procedures he effected with the air-pump, or of increased pressure, as for example the 'Bone Digester' of Denis Papin, a kind of pressure cooker, which was a forerunner of the steam-engine.[6]

In traditional understandings, the air was ubiquitous but essentially plural. It is perhaps for this reason that the air has not been conceptualised as the other elements have in many cultures and systems of thought. Hebrew had no single, all-encompassing word for 'air', in all its aspects.[7] In this it resembles Sumerian. Although there is a Sumerian god of the air, Enlil, there is no term for the air at rest in Sumerian, or in Akkadian, all the terms in fact denoting various kinds of motion – wind, storm, etc.[8] Air is not to be found among the elements in the earliest Indian texts; the Chandogya Upanishad (eighth or ninth century BCE) limited the elements to three: fire (*agni*), water (*ap*) and earth (*prithivi*). Nor does air feature among the five elements, wood, fire, earth, metal and water, discriminated by classic Chinese texts such as the Tao Te Ching (sixth century BCE). The air was clearly identified as an element by pre-Socratic philosophers. Indeed, Anaximenes (sixth century BCE) made it the original out of which all matter was made, being followed in this by Diogenes of Apollonia. Nevertheless, there was in Greek thought a strong tendency to see the air as multiple rather than singular in form. Indeed, its conception of the air was, as it were, distributed between two words: the *aether*, the realm of brightness above the clouds, and *aer*, the dim, moist atmosphere below them. More than any other element, the air's essence is to be dissimulated into its aspects, accidents, and appearances – as breath, wind, height, space, transparency, lightness, light. This essential plurality of the air explains why Aristotle could be so emphatic in the arguments in his *Meteorology* against what seemed to him to be the crackpot notion that all winds were effects of the disturbance in the air:

> Some say that what is called air, when it is in motion and flows, is wind, and that this same air when it condenses again becomes cloud and water, implying that the nature of wind and water is the same. So they define wind as a motion of the air. Hence some, wishing to say a clever thing, assert that all the winds are one wind, because the air that moves is in fact all of it one and the same; they maintain that the winds appear to differ owing to the region from which the air may happen to flow on each occasion, but really do not differ at all. This is just like thinking that all rivers are one and the same river, and the ordinary unscientific view is better than a scientific theory like this. If all rivers flow from one source, and the same is true in the case of the winds, there might be some truth in this theory; but if it is no more true in the one case than in the other, this ingenious idea is plainly false.[9]

The air manifested itself in a dizzying and seemingly irreducible plurality of forms. Maurice Crosland has written:

> When the full early history of gases comes to be written, it will include chapters on mineral waters, mines, caves, and other natural conditions. If one were making a study of vocabulary, something would have to be said about 'exhalations,' 'smokes,' 'steams,' 'damps,' 'vapors,' 'mofettes,' 'spirits,' and so on.[10]

The grasping of the air's nature was blocked by one simple fact; despite being everywhere to hand, and nowhere absent, the air could not easily be isolated. In order for the universal and invariant features of air to be understood, the air had to be partitioned. In order to be seen and known as itself, air had to be divided from itself, deprived of its principal features, namely its tendency to mix, compound and diffuse, along with its ubiquity. How does one study a substance that is everywhere? Normally, an object of study is something that is, following the etymology of the word, 'thrown before', or set out in front of the investigator. To study an object, one must pick it out from its surroundings, and concentrate it in one place. How was one to make of the air such an object? How was the air to be picked out of its surroundings, when air was ambience itself? How was the air to be brought before one, when it was of necessity and at all times all about? An object, a body, as it might more commonly have been called in the seventeenth century, must exist in a space of observation. How can the substance that seems to have the function of filling space, and so seems to approximate to space itself, itself be set out in space? This would mean trying to isolate it from itself. Since the air was everywhere in the world it was coextensive with it, and trying to see round the air was like trying to see round the world. Until one could travel out of the world, and therefore out of the air (and people began to wonder more and more concretely about this possibility during the 1600s), it seemed that such a vantage point would never be attainable.

But there was another way. What if, rather than trying to roll the air up into a ball that one could look at from the outside, one were to produce a space of observation – an air-lock – within the very space of the air? What if one were to try to give that impossible, out-of-this-world vantage point a local habitation and a name inside this world? What, in other words, if one were to open up a space in the world in which there was no air, and therefore no world? Aristotelian science and theology agreed that this was an impossibility, because space was a plenum. There could be no void places within nature, for nature would not tolerate it. The technology to investigate this possibility practically lay to hand, and had been to hand for centuries, for it was known that it was possible to draw water uphill by withdrawing a piston in a sealed chamber. Galileo held to the orthodox view that this was an effect of suction, namely that the attempt to create a vacuum drew the water in after it. The problem with such an argument was that the repugnance of nature towards a vacuum seemed to have a limit. No matter how one tried, one could not 'pull' up a column of water at any one time greater than about 5 m. When, in 1644, Evangelista Torricelli, who had become interested in the ageing Galileo's work on the nature of air, filled a one-metre tube with mercury and then inverted it in a bath of mercury, the level of mercury in the tube fell, leaving a gap at the top. This was not the first time that such a loophole in space, and apparent suspension of the law of nature, had appeared. It was not producing a vacuum that presented the difficulty, but acknowledging and explaining it. Alternative explanations rushed in. Franciscus Linus argued that the mercury was held up by a narrow fibre of subtle mercury he called a 'funiculus'. What the 'suckers' like Galileo and Linus maintained, against the 'shovers' like Torricelli and, later, Robert Boyle, was that the mercury was pulled up by the outraged near-vacuum, rather than pushed up by the weight of the air outside. As Boyle put it, protesting in his *Defence of the Doctrine Touching the Spring and Weight of the Air* (1662) against the doctrine of the 'strange imaginary Funiculus': 'whereas we ascribe to the Air a Motion of Restitution outwards, he [Linus] attributes to it the like Motion inwards'.[11]

All the most important experiments involved the effort to create enclosures in which states of exception from nature, or nature in a state of exception from itself, could exist and be observed. This depended upon the capacity to make apparatus that would effect it. The universality of the air could only be understood if that universality could be set aside or suspended – hence the singular importance for the understanding of the air of equipment that produced and sustained states of vacuum.

There was one particular aspect of this equipment to which Boyle's detailed accounts again and again draw attention, namely the fact that, in order for the effects of extinguishing candles, firing pistols and suffocating birds to be observed, the air must be sequestered without being secreted, made divisible while yet remaining visible. Only one material possessed the requisite qualities of impermeability and perspicuity, namely glass. Only glass allowed the air to be kept both apart and apparent, bringing different worlds into close proximity. In a sense, glass revealed a kind of mystical confederacy with the air it enclosed and displayed. Like the air, it had spring and ponderability, yet seemed invisible. Since the material conditions of the glass that Boyle employed established the conditions for and set limits to the questions that he could ask of the air, one might almost imagine that Boyle was working on a composite matter, a kind of vitreous air, or gaseous glass. The fact that the glass was in the form of globes, bubbles and spheres was reminiscent of the crystal spheres that had compounded the ideas

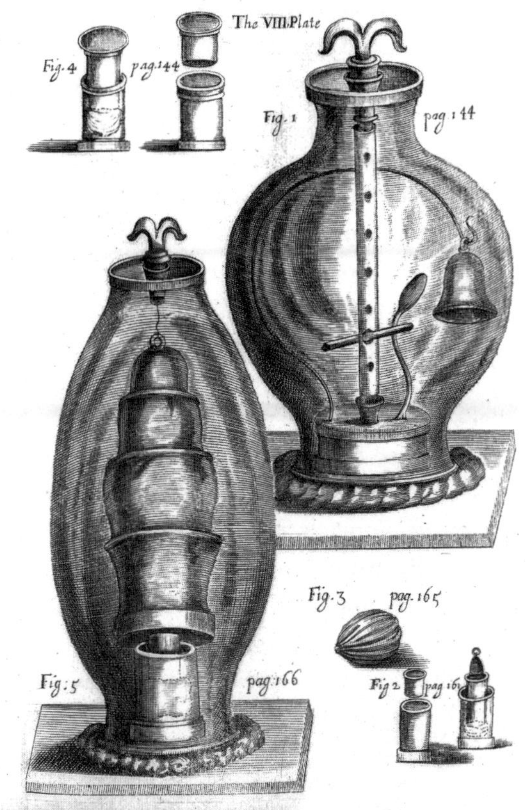

Experiments with Robert Boyle's air-pump: (left) an expanding bladder of air lifts a pile of weights; (right) ringing a bell to demonstrate the absence of sound in a vacuum. Robert Boyle, *A Continuation of New Experiments Physico-Mechanical, Touching the Spring and Weight of the Air and their Effects* (Oxford, 1669).

of glass and air in the Ptolemaic cosmology. But there was an even closer and more intimate affinity between glass and air, in the fact that those globes were formed through a blowing process that installed the glassmaker's breath at the heart of the structure, making air a constitutive principle of the glass. The globe blown by the workman was a manifestation of the very principle of aerial spring that it was designed to display. The many enclosures of the air, artificial atmospheres and sealed environments that would follow in the centuries after Boyle's researches – crystal palaces, cockpits, arcades, greenhouses, light-bulbs, diving helmets and pleasure-domes – would continue to be brought about by the favoured material of glass.

Not that glass was the only material necessary for Boyle's researches. He made frequent visits to butchers to obtain moist, elastic bladders, lamb's bladders seeming to be particularly favoured. One of his most impressive demonstrations of the expansive power of the air involved putting a sealed bladder containing a small amount of air into a sealed container. When this was evacuated of air, the air inside the bladder expanded, making it swell like a balloon (*Works*, I, 174). Imaginary matter was as important as actual matter in providing dramatisations of the force of the air. The metaphor to which Boyle returns throughout his first discussion of the spring of the air is of a mass of wool, which, first compressed in a closed hand, then released, will spring back to its original volume (*Works*, I, 165). This metaphorical wool comes to cooperate closely with the other actual substances. One must add to this the various pastes, oils and other materials that Boyle used to seal his apparatus and attempt to prolong the lifetime of the vacuum inside it for as long as possible, while proportionately shortening the lifespans of the creatures he experimentally introduced into it. He even finds a way to recruit the air itself to this sealing function. On one occasion, he advises the reader attempting to replicate his experiments not to assume that glass containers that have cracked as a result of the force of the air expanding inside them are necessarily without further use, since sometimes the very force of the air pressing on the outside of the cracked glass will itself push the sides of the crack together, the air thereby helping to form a seal against itself (*Works*, I, 183).

The principle of enclosure encouraged Boyle to experiment with doubled, or interior enclosures. The most celebrated of these involved putting inside his vacuum chamber the entire set-up used by Torricelli to show that the pressure of the atmosphere pushes up a column of mercury. Now the air in his chamber would constitute, not an enclosure, but an outside with respect to the volume of mercury on which it was exerting pressure. When the air was drawn out of the chamber, this pressure was reduced and so the column of mercury fell. This demonstrated clearly that the column of mercury was pushed up by the pressure of the atmosphere, and not pulled up by some occult quality of the vacuum at the top of the column, as some (most notably the great Galileo) thought. Boyle's experimental apparatus was a theatre of inversions, in which inner and outer exchanged positions promiscuously.

This play of inside and outside has larger implications. Bernadette Bensaude-Vincent and Isabelle Stengers note that the principal difference between the alchemical tradition and the emerging scientific tradition was not so much a difference in theory and even less a difference in apparatus and technique, for the alchemists developed to a high degree many of the techniques that would continue to be important in chemistry. Rather, it was a difference in the conditions under which chemical investigation was conducted, as enabled and typified by the spread of printing.[12] Where alchemy was an esoteric knowledge, a Hermetic knowledge, sealed, encrypted and enclosed in manuscripts laboriously copied and kept hidden from the eyes of the vulgar, printing allowed and demanded an opening out, an exposure to view, of scientific investigation. This investigation became more and more a matter of technique and technology rather than of wisdom or secret inspiration. It made no difference that the chemist was himself a refined spirit. The chemists of the early seventeenth century were frequently called 'virtuosi', meaning savants, specialists or learned persons. By the middle of the eighteenth century this usage had become obsolete and the word was beginning its association with the arts, especially the arts of music. Robert Boyle's voluminous writings exemplify this new spirit of explicitness and openness. The point for Boyle, and the reason that he is often regarded as the first chemical scientist, was to lay his experimental procedures open to view, largely in order that they might be repeatable by others.

Boyle was a member of what was known as the 'Invisible College', a group of investigators who were the forerunners of the Royal Society. They included John Wilkins, Robert Hooke, Christopher Wren, William Petty and Thomas Willis. They were not invisible in the old-fashioned sense that they met secretly, or at least privily. What Boyle meant when referring to the Invisible College in his letters was the fact that they had no building or established institution in which to meet. Though there were meetings of those members of the society – and the precise membership is hard to establish[13] – the society was dispersed, and therefore to some degree invisible to themselves as well as to others. Like a present-day network, they were constituted through the exchange of letters, manuscripts and objects. Boyle's first mention of the term 'Invisible College', which he perhaps originated, is in a letter to Isaac Marcombes of 22 October 1646.[14] In his second use of the term, in a letter of 20 February 1647 to Francis Tallents, Boyle writes jestingly that 'the corner-stones of the *invisible*, or (as they term themselves) the *philosophical college*, do now and then honour me with their company'.[15] His next use of the term is in a letter of 8 May 1647 to Samuel Hartlib, in which he seems to complain that Hartlib is too taken up in the virtual life of the Invisible College: 'you interest yourself so much in the *Invisible College*, and that whole society is so highly concerned in all the accidents of your life, that you can send me no intelligence of your own affairs, that does not (at least relationally) assume the nature of *Utopian*'.[16] The Invisible College was therefore invisible, not because it was closed, but because, like the air, it was open.

The word 'hermetic' underwent something of the transformation that would be undergone by the word 'pneumatic'. The primary reference of the word is to the person or practice of Hermes Trismegistus, the mythical founder of alchemy. The word retains its mystical tincture, for example in the use of the term by Stephen Hales to characterize the gaseous (not his term) condition itself, in suggesting that

> If those who unhappily spent their time and substance in search after an imaginary production, that was to reduce all things to gold, had, instead of that fruitless pursuit, bestowed their labour in searching after this much neglected volatile *Hermes*, who has so often escaped thro' their burst receivers, in the disguise of a subtile spirit, a meer flatulent explosive matter; they would then instead of reaping vanity, have found their researches rewarded with very considerable and useful discoveries.[17]

But during the seventeenth century, the esoteric use of the word was in competition with a much more workaday meaning. From the 1660s the term 'hermetic' or 'hermetical' referred normally to the techniques employed to prohibit the entrance of air into sealed vessels, often by melting the neck of a glass flask or bottle. The *Oxford English Dictionary* gives the first appearance of the use of the word to refer to the 'air tight closure of a vessel, esp. a glass vessel, by fusion, soldering, or welding' as Jeremy Taylor, in 1663. Taylor uses the term as a metaphor for the stopping up of change of the immortal soul after the Resurrection:

> [A]s *Jesus Christ is the same yesterday and to day, and the same for ever*, so may we in Christ become the morrow of the Resurrection, the same or better than yesterday in our natural life; the same body and the same soul tyed together in the same essential union, with this only difference, that not Nature but Grace and Glory with an Hermetick seal give us a new signature, whereby we shall no more be changed, but like unto Christ our head we shall become the same for ever.[18]

But William Oughtred precedes him in 1653, in the course of advising on a procedure for driving off vapour from nitre which requires a glass vessel that is closed 'hermetically'.[19] Robert Boyle also uses the term frequently. Although there is a reference to Paracelsus as 'our Hermetick Philosopher (as his followers would have him stil'd)', nearly all his uses of the word refer to various modes of physically sealing vessels from the air (*Works*, I, 287). Thereafter, the word appears frequently in Boyle's works, most particularly in his *New Experiments and Observations Touching Cold* (1665), in which he refers to his thermometer as a 'Hermetical Thermoscope (if I may for distinction sake so call It, by reason of its being Hermetically seal'd)' (*Works*, IV, 237).

Although Boyle refers to the unusual tools and dexterity required to produce these sealed vessels, the point of his enclosure of the air is precisely to enable that 'assiduous Conversation with Nature' (*Works*, IV, 211), which the open air does not permit. It is the open air that is secret, withdrawn, folded upon itself: sequestered, partitioned, in bell-jars, tubes, flasks, chambers and bladders, the air can be made to open up.

A little later in the century the word had been so trafficked about as to be used, with only a hint of irony, in a communication to the Royal Society regarding 'The Anatomy of a Monstrous PIG', in which unfortunate animal 'the end of the *Rectum* was entirely closed like a bladder, and sealed as it were Hermetically'.[20] The word was also used

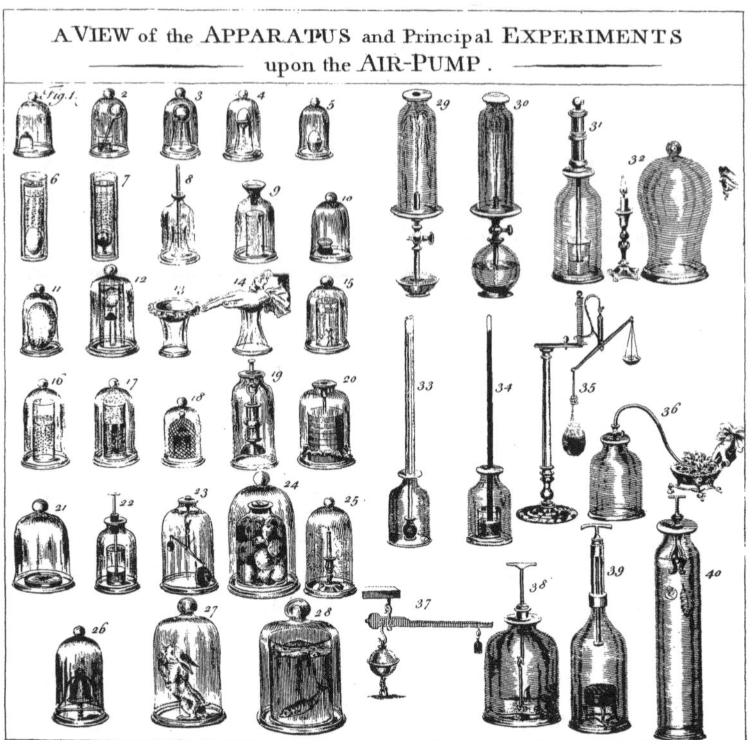

Air-pump apparatus and experiments, from Benjamin Martin, *The Description and Use of a New, Portable, Table Air-Pump and Condensing Engine. With a Select Variety of Capital Experiments* (London, 1766).

of an instrument called the 'aeolopile', which was defined in Edward Phillips's *New World of English Words* in 1658 as 'a kind of Instrument called the Hermetical bellows, whereby it is experimented whether there be a *vacuum in nature*'.[21] None of these usages suggests that the word was outlandish or in need of explication, except perhaps that of Henry Power, who specifies in a list of required equipment in *Experimental Philosophy* of 1664 'Several Glass-Trunks, or Cylindrical Glass-Tubes, some open at both ends, and some exactly closed; or (as they phrase it) Hermetically sealed at the one end'.[22]

Gradually the esoteric provenance of the word was concealed by the demotion of the capital letter: 'Hermetic', became simply 'hermetic'. By 1675 the term was current enough for it to be used in a book of recipes for foods and cosmetics. Hannah Woolley's *The Accomplish'd Lady's Delight in Preserving, Physick, Beautifying, and Cookery* advised the following technique for making spirit of ambergris: 'Take of Ambergrease two drams, of Musk a dram cut them small, and put them into a pint of the best spirit of Wine, close up the glass Hermetically, and digest them in a very gentle heat till you perceive they are dissolved'.[23]

The association of the new pneumatic science with the power to enclose the air is lampooned tellingly in Thomas Shadwell's satirical *The Virtuoso*, in which Sir Nicholas boasts of his collection of different airs, as another might laud his wine cellar:

> *Sir Nic.*: Chuse your Air, you shall have it in my Chamber; *Newmarket, Banstead-down, Wiltshire, Bury*-air, *Norwich*-air; what you will.
>
> *Bruce*: Would a man think it possible for a Virtuoso to arrive at this extravagance?
>
> *Longv.*: Yes, I assure you; it is beyond the wit of man to invent such extravagant things for them, as their folly finds out for themselves. Is it possible to take all these several Countrey Airs in your Chamber?
>
> *Sir Nic.*: I knew you were to seek. I employ men all over *England*, Factors for Air, who bottle up Air, and weigh it in all places, sealing the Bottles Hermetically: they send me Loads from all places. That Vault is full of Countrey-air.
>
> *Bruce*: To weigh Air, and send it to you!
>
> *Sir Nic.*: O yes, I have sent one to weigh Air at the Picque of *Teneriff*, that's the lightest Air. I shall have a considerable Cargo of that Air. *Sheerness* and the Isle of *Dogs* Air is the heaviest. Now if I have a mind to take Countrey Air, I send for, may be, forty Gallons of *Bury* Air, shut all my windows and doors close, and let it fly in my Chamber.[24]

In fact, however, the tendency of the kinetic experiments with the air undertaken by Boyle and his followers was to establish the universal conditions of the air. Ironically, Boyle himself believed that air was full of effluvia. Nevertheless, his concern was with what, in his

General History of the Air (1692), he called '*the constant and permanent* ingredients *of the Air*' (*Works*, XII, 12). He explicitly says that he means by the air 'not, either the pure Element of Air, which *some*, nor that Etherial or Celestial Substance, that *others* . . . assert' (*Works*, XII, 12).

Ramification

The phase of what Boyle called 'physico-mechanical' investigation, or investigation into the powers and effects of the air, lasted until the middle of the eighteenth century. It was not until this period that real advances were made in the understanding of the chemistry, or the composition of the air. This seemed to ask a different kind of question of the air. Mechanical investigation had found, in the production of states of vacuum, a solution to the problem of how to put the air into abeyance, suspending its operations in order to see it, as it were from the outside, as an outside that was paradoxically enclosed conveniently in visible and manipulable interiors. The mechanical question asked: 'how can one get outside the outside, or put that exteriorized outside inside? The chemical question by contrast was: 'how can one get inside the air?' Everything is in the air, but what lies on its inside? Might it be possible to extract this essence, bringing the inside principle out from its interior? The investigation of the air was made possible by the sequestering powers of the air-pump, which allowed the air to be subtracted, or set aside from itself. The chemical investigation of the air would depend upon another artifice, namely the production of 'factitious airs', artificial variations in the form of air, that would eventually reveal the compound nature of air in the first place.

This investigation begins with the identification of aeriform fluids that were not air. Although gaseous effluvia, in mines, marshes and volcanoes, and as a result of industrial processes such as brewing and dying, had been known for centuries, they were thought of as altered or vitiated states of air. The ancient world was familiar, indeed, preoccupied to the point of obsession with, various kinds of vapour, exhalation, fume and spirit, most of which we would nowadays characterize as aerosols or similar suspensions of liquid or particulate matter in air. But hardly anything was known of the more mysterious kind of aeriform substance that was occasionally discovered or arose in chemical experiment, which, because of the difficulty of capturing, controlling and condensing it, was known simply as 'spiritus silvestris', wild spirit, or 'incoercible spirit'.[25] That inflammable gas occurred naturally in mines (in the form of deadly firedamp), and issued from swamps and clefts in the earth elsewhere, had been known for thousands of years. Only in the seventeenth century do reports begin to appear of an inflammable gas that could be produced as a by-product of the process of coking, or distillation of coal. Paracelsus was aware of this possibility, but it fell to Jan Baptiste van Helmont, who acknowledged the work of Paracelsus, but averred that 'that man was ignorant of the thingliness of a Gas', to give the substance a more specific name:

> by the Licence of a Paradox, for want of a name, I have called that vapour, Gas, being not far severed from the Chaos of the Ancients. In the mean time, it is sufficient for me to know that Gas, is a far more subtile or fine thing than a vapour, mist or distilled Oylinesses, though as yet, it be many times thicker than air . . . [T]he live coal, and generally whatsoever bodies do not immediately depart into water, nor yet are fixed, do necessarily belch forth a wild spirit or breath . . . I call this Spirit, unknown hitherto, by the new name of Gas, which can neither be constrained by Vessels, nor reduced into a visible body.[26]

Despite Van Helmont's explicit reference to the Greek 'chaos', Antoine Lavoisier thought gas derived from Dutch *ghoast* (spirit), while the German physician Johann Juncker defended Van Helmont against those who thought his word 'barbaric', by claiming that it came from German *gascht*, or *gast*, ferment, froth or foam.[27]

Van Helmont could not see how such a strange and paradoxical substance could already be in the coal that he roasted or the grapes he boiled.

> Bodies do contain this Spirit, and do sometimes wholly depart into such a Spirit, not indeed, because it is actually in those very bodies

> (for truly it could not be detained, yea the whole, composed body should flie away at once) but it is a Spirit grown together, coagulated after the manner of a body, and is stirred up by an attained ferment.[28]

Van Helmont coined a companion term, 'blas' (suggested presumably by the word 'blast'), to name a physical influence exerted by the stars, in the form of a wind or flatus. Though frequently associated with gas, this term had less success and indeed its contaminating assonance probably delayed the acceptance of 'gas'. The word 'gas' entered only slowly into use among English chemical writers within a couple of decades. Boyle uses it, and the physician George Thomson, a follower of Van Helmont, was keen on the word, which he defined as 'a wild invisible Spirit, not to be imprisoned or pent up, without damage of what conteins it, arising from the Fermentation of the Concourse of some Bodies, as it were eructating or rasping this untamable Matter'.[29] Thomson, who tells us that he spent some hazardous years investigating the 'treacherous Gas of Charcole',[30] even assigned the 'material cause' of the plague, with impressive circumstance, to

> a venemous Gas, or wild Spirit, produced either inwardly from some degenerate Matter conceived within the Body, or outwardly received from some fracedinous noisom Exhalations contained in the pores of the Air, taking their Original from several putrid bodies excited to fermentation, rarified and opened by the Ambient, altered and moved by Celestial Influences, and so disposed to this or that condition, whereby an expiration is made of virulent Atoms, which sometimes close pent up, and drawn in by the Lungs, do by their deletery power put to flight, confound, suffocate and mortifie the Archeus, or vital spirits, in as short a times as the strong fume of burning Brimstone doth destroy any small living Insect held over it.[31]

But Walter Charleton was still referring snootily in 1669 to 'the Gas Sylvestre (as Helmont barbarously calls it)'.[32] Samuel Parker referred offhandedly to 'Gas and Blas or any other words of no defined signification' and, mocking the new-fangled word 'barometer', Henry Stubbe sneered that 'they had done better to have called it . . . Gas, or Blas, or Diaceltateston, or some such unintelligible name'.[33] John Norris was still complaining in 1697 about the tendency of men to believe the mysterious rather than the simple, giving their credence to those who account for the effects of nature

> by the yet more Obscure Principles of the Chymists, striking and filling their Ears with those great but empty Sounds, *Archeus, Seminal Spirit, Astral Beings, Gas, Blas,* &c. which they receive with great satisfaction not for their Scientific Light (for they are dark as may be, mere Philosophic Cant) but only because they are Mysterious and Abstruse.[34]

Boyle had suspected that there was a principle hidden in the air that his experiments had not been able to isolate or bring into view. John Mayow is often credited with identifying the active principle of the air. Mayow had shown in 1674 that an enclosed volume of air was reduced both by the combustion of substances and the respiration of an animal. Furthermore, he showed that the remaining air would no longer support combustion, indicating clearly that some active principle in the air, which Mayow thought consisted of 'nitro-aerial particles' or 'aerial nitre', was involved in both combustion and respiration.[35] This gave experimental confirmation to a hypothesis that was shared among other investigators, including Kenelm Digby, Robert Hooke Thomas Willis and even perhaps, considerably earlier, Paracelsus.[36]

Little was done to follow up Van Helmont's work with artificially produced air for almost a century. Even then, there was a seemingly unshakeable assumption that the products of the processes employed by Black, Cavendish and Priestley, and the subjects of their experiments, were still variants of air.

If the mechanical investigations of the first generation of modern air-science established the properties of the air as a body, the real history of the air begins with the dissolution of that body, following the discovery that air was not a single substance at all. Pneumatic chemistry takes its starting point from the primal splitting of air, or the discovery that air was not the only elastic fluid, indeed that air itself was made up of different constituents. In one sense, the domain of the air shrinks remarkably, as it is identified as just one compound of gases amid many other actual and possible. In another sense, the air expands and ramifies, as 'the air' comes to typify and be identified, not just with a particular substance, but with the gaseous state of matter as such. After its brief parenthesis, during which it was isolated and universalised, the air resumed its mongrel, combinatory condition.

Next Degree to Nothing

Reflections on air produce complex convolutions, even convulsions, of inner and outer, as we learn to get on the outside of our own interiority with respect to the air, but also draw out from the inside of that outside principles, processes and practices. The mechanical phase of the air's investigation concentrated on the air's outer effects, even though it had to invent and construct new world-spaces, new pockets of exception in the immersive totality, in order to see them. The chemical investigation of the air focused on the air's inner nature, and, given the centrality of oxygen and the new understanding of the action of respiration to this phase of pneumatic chemistry, on the inner workings of air in human physiology At each turn, more crossings and chiasmas, more reversals and traversals of the relations between inside and outside become possible.

There is a further convolution in this spiral that implicates the action of thinking of the air in the nature of air. The air is unlike the other three elements in one particular respect. For, as Robert Boyle wrote in his *General History of the Air*, 'the Generality of Men are so accustomed to judg of things by their Senses, that not finding the Air to be a visible Body, they ascribe less to it than even the School-men do; and what is invisible, they think to be next Degree to nothing' (*Works*, XII, p. 132). All the other elements also present themselves under many different forms – water, as the root of many different kinds of liquid and their differing states of liquidity, earth as the collective name for all the forms of solid, shapeable, resistant matter, and fire as the name for the fusing and dissolving effects of heat. But air is larger in its scope than the other elements because it encompasses its own negation, indeed perhaps even negativity itself. Take away the air, and the empty space you have left still seems to retain most of the qualities of air. The air is unique among the elements in having this affinity with nothingness, in signifying the being of non-being, the matter of the immaterial.

The air is also paradoxical, because it encompasses so many opposites and antinomies. Unlike earth, water and fire, it is imagined in at least two different forms; as a lower, so to speak, earthly air, the air that is breathed, that blows in the form of wind; and the higher air, the ether. The real air is, so to speak, always accompanied by an ideal or phantasmal air. No other element is divided in this way. Air is the thing that is nothing, the unbeing that is.

As such, it is not just next degree to nothing, but next door to the peculiar kind of nothing that is the space of thought. The convolvement of air with the thought of air is another form of our implicatedness in the air. This is both older and younger than the chemical revolution, or the scientific revolution. Air has traditionally been thought of as the favoured element or state of matter through which to body forth the thought of thought. As air became more and more materialised, so it became more and more the body of thought. But, from being an image of the immediacy of thought, the air became an image of the mediate matter of thought. The air became ever more necessary to thought as it became in later centuries the arena of angelic hosts of other mediate matters, electricity, radiophony, radioactivity.

Thus, the air has become the element, not of our exposure or containment, but of our immixture with, our inextricability from the outside world. The air has become mediation itself. We find our own image in this mediate matter (mediated and mediating), not just because we communicate through our (largely airborne) media, but because we are the throughput of those mediations. Getting outside ourselves, in the ultimate, ecstatic form of what Peter Sloterdijk calls 'explicitation' (*Schäume*, 87), we recognise and construct our immersion and thus, leaving and resuming it at once, we 'immerge' with the world, emerge into a knowledge of our immersion, in which we are both inside and outside our previously implicit relation and inside and outside our knowledge of it.

Elaboration

This book explores a number of arenas of this immergence, areas in which the air has been newly estranged, extrinsicated, applied, multiplied, complicated. It will be concerned with the ways in which the air was *put to work*. But each of these areas draws us in, becomes a mirror or domain of our new condition. We renewedly renature the air, but each time further volatilise our own natures. We become ourselves more and more what we make of air. In taking air for ourselves, we take more and more to it.

Chemistry is to be distinguished from other sciences in two respects. The first is that is has a profligate adolescence, from which it must make a break. This history bears the name of alchemy. Chemistry comes into being in its determination to cast off the imposture and vanity of alchemy. Although other sciences have their histories of limits surpassed and errors corrected, no other science has the same relation to a specific prehistory of radical error. The understanding of the air had a key part to play in this purification of the science of chemistry. Where alchemy mixed up spiritual and material, the air, the carrier of the spirit, was drawn across into the side of materiality.

But the other distinguishing feature of chemistry is that it is itself a mixed body, a compound of different practices:

> It is astonishing that techniques as dissimilar as metallurgy, pharmacy, and the art of the perfumer were ever assembled under a common denominator and melded into the common territory of 'chemistry', with its own culture, methods, and identity.[37]

For this very reason, chemistry has remained an impure science, a mixed body. Chemistry has in fact never quite succeeded in detaching itself from applications of all kinds, cosmetic, medical, military, industrial in the broadest sense. Chemistry has remained worldly, has remained a kind of working on the world.

One of the crucial features of the new science was the importance of equipment and apparatus. As Robert Hooke observed of an improved lamp of his own devising,

> For this I look upon as one of the Tools to be made use of in the Work-house or Elaboratory of Nature, without a good Apparatus of which, be the Workman otherwise never so well accomplished, he will never be able to produce any very considerable effect; and with them, even a Bungler otherwise, will, if well furnished, do wonders to such as know not the means by which they are done.[38]

Equipment helped, not so much in the objectifying of nature as in the desubjectifying of science, which are not at all the same thing. When the subject of scientific investigation is no longer a crafty magus but a workman, nature itself is able to partake of the process of elaboration (hence perhaps the interesting ambivalence of that 'of' in the phrase 'elaboratory of nature', which allows us to understand nature as implicated in its own explication). Indeed, during the seventeenth century, the metaphor of nature as itself containing or consisting of 'elaboratories' was common, especially in accounts of the workings of the human body. John Howe reverently described the process whereby the blood was conveyed 'upwards, to its admirable elaboratory, the heart: where it is refined and furnished with fresh vital spirits, and so transmitted thence by the distinct vessels prepared for this purpose'.[39] For Robert Johnson, the stomach rather than the heart was the body's principal 'elaboratory':

> The Stomach (in continual Fevers) is most commonly primarily affected through undigestion, or else from Excrements, not being separated and orderly evacuated; which causeth an irregular Ferment, or nonnatural heat in the Stomach; which (though begun else where) is much aggravated by vitiating Juices, found in this first Elaboratory of decocting Nature.[40]

Athanasius Kircher imagines volcanoes as part of a system, in which Vulcan has 'his Elaboratories, Shops, and Forges in the profoundest Bowels of Nature'.[41]

Bensaude-Vincent and Stengers see Van Helmont's discoveries of gas as a bridge between the old and new chemistry. They point out that, although Van Helmont broke with many alchemical ideas, believing in particular that there was only one primary substance, water, from which all other substances can be formed, he continued to believe that minerals and other material substances were generated by the same processes as those that generated living forms. This would require, not only a primary substance (water), but also what Paracelsus described as the *archeus*, that which gave shape and form to substances. 'The question then is: through what material intermediaries, or by what instruments, does this seed – or the *archeus*, the agent postulated by Paracelsus to give form and unity to natural bodies – operate?'[42] They suggest that Van Helmont saw the gas that he named and investigated as the carrier of this principle: 'he identified "spiritus sylvestris" (our carbon dioxide), which emerges when a natural substance, which has a shape, is destroyed by burning, as when acid is poured on limestone or during fermentation.'[43]

While I believe that distinctive and far-reaching changes have taken place in the human relationship to the air over the last four centuries, readers expecting to be able to take away from this book either an account of the sempiternal meaning of the air for humans, or a once-and-for-all story about the ways in which the human experience of the air has changed, may be made irritable. While insisting that human history visibly moves in the direction of softness, lightness and volatility, Michel Serres has also argued that the 'there' of human 'being-there' has always been more essentially air than earth. In an interview of 1995, he said that: 'we are the dasein in the sky, not in the land. Do you see what I mean? We are wandering. We are nomads. This is not a new state of things. It is a very ancient state of things. I think the dasein is in the atmosphere.'[44] So, rather than a sudden tearing of tectonic plates, or wrenching of the ground (the kind of *Luftbeben* or 'airquake' that Peter Sloterdijk tries to make out, *Schäume*, 89), which gashes open a permanent crevasse, from the opposite sides of which the ancients and moderns can only peer and gesture perplexedly at each other, new understandings like that of the air dissolve and pervade, and in the process form both mixtures and compounds with previous but still persisting ways of thinking and being. Gases propagate. Ideal gases under ideal conditions would propagate at an absolutely even rate, spreading in all directions equally fast and eventually becoming imperceptible. But we are not interested in ideal conditions, nor, it seems, are they in us, and in actual fact gases will tend to aggregate and linger, interacting in specific ways with existing components of the atmosphere into which they are released.

The form of this relation is not paradoxical, or only from the viewpoint of dry land. Rather, it is helical, vortical, characterized not by fixed positions but by the speed and force of the process whereby positions change relative to each other. So we should not find it strange that the air should resist exclusive thought, or that it should require thinking in terms of the ways it itself inverts and involves the inside and the outside, the old and the new, the natural and the human.

The air has often carried the idea of the infinite, the illimitable, the transcendent. This notion depends upon a sense of the air not as an element, or as a substance, but as a dimension. The dimensionality of air is enlarged in the work of Gaston Bachelard and Luce Irigaray, the two writers who have most extensively considered the role of the air in the poetic and philosophical imagination. For both these writers, the air has the meaning of infinitude. For Bachelard, the air stands for the lightness and freedom of aspiration; it is really a metonymy for the power of flight that takes place through it.[45] Nietzsche, to whom Bachelard devotes a whole chapter, is the airiest philosopher, because of his love of height and eminence, of the dimension of the *über*, or overness. Irigaray takes Heidegger to task for what she calls his forgetting of air, by which she means, not so much height, as breadth. Air means that which goes beyond. Where Bachelard finds in poetic dreams of air an identification with the element that promises escape and enlargement (air means the flight that takes place in and through it – air itself is flight, or in flight), Irigaray emphasises the transcendence of air, which must always exceed any attempt at containment or identification – an identification *with* air that would be accessory to an identification *of* it. Air is not a place, or a substance for Bachelard and Irigaray: it is a direction, an expansion, a multivectorial potential.

For many writers, the air signifies the wholly open, whether in the mode of yearning dream (Bachelard), or melancholy regret (Irigaray), or of impassioned possibility (Serres). But this book will not be able to approve any of these forms of ecstatic agoraphilia, either as a description of our condition or as a model for our aspiration. Instead, it will work through some episodes and arenas in which the relations between the sense of the air as open and the enclosure of the air both alternate and form new complications. For we are both more open, more porous, more pervaded, more available to each other, less gathered together in one place and time than ever before, and yet also more hived, more isolated, more reclusive, more encrypted. Indeed, the password is the contemporary form of our open secrecy: the means by which we gain and constrain access to a fine and private place within the otherwise open airspace of public communications. Our contemporary moment is characterized by the complex imbricated topology of these two alternatives, operating at all levels – information, architecture, disease, travel, communication. Peter Sloterdijk's 'foam' is meant to image this ambivalence, but perhaps the best example is the human lung. Unlike the lungs of frogs and other amphibians, which tend to be simple bag-like structures, with the area of gas exchange limited to the inner surface of the lung, the lungs of mammals like human beings provide for their increased energy demands by around 300 million bubble-like structures called *alveoli* (Latin, little alcoves), which compress into the volume of the lungs a total surface area of around 75 square metres. The lungs are neither open nor closed: secluded in the body, they are nevertheless the body's form of exposure to its outside, its way of creating a field of openness within its own compass.

Perhaps the ultimate form of this ambivalence concerns our new relation to the atmosphere. In one sense, given our new recognition, not just of our susceptibility to the atmosphere, but also of our power to affect it, we have hemmed the atmosphere in, getting outside our own outside. In another sense, we have created a new kind of exposure, to unpredictability or run-away consequences. We are exposed to the finitude that we have made of the air.

The chapters that follow treat the different mutations that the idea of the air has undergone over the last two centuries, following upon the mechanical and chemical explications of air and the gaseous state. These discussions do not follow in an orderly or obvious way one upon another, though they drift from the beginning of the nineteenth century to the concerns with air and atmosphere of contemporary artists.

There are three partitions, or they might better be described as phases, in the book. Part One explores some of the ways in which air, and the thought of the aerial, were put to work during the nineteenth century. As so often, delusion works in tandem with and perhaps even somewhat ahead of the work of technicians and industrialists. Accordingly, its first chapter considers in some detail some examples of imaginary machines or machineries in which air or gas are important elements, centring on James Tilly Matthews's systematic delusions of a pneumatic influencing machine. Matthews's delusions will be shown in oblique but intimate relation to the new pneumatic sciences that established themselves in the final decades of the eighteenth century. The next chapter, 'Inebriate of Air', considers the coming together of air and thought in a literal kind of way, through an account of William James's philosophical transactions with nitrous oxide. Chapter Four, 'Gasworks', then moves from imaginary to actual machineries, to consider a neglected area of nineteenth-century technology, the development of gas lighting, and the ways in which it changed conceptions of the matter of air, making actual in a range of new ways its condition as a 'mediate matter'. The final chapter in this part of the book, 'Transported Shiver of Bodies', finds a way of thinking through the most massive Victorian thought-experiment with air, in its long preoccupation with the idea of the ether. This chapter brings to explicitness a feature of the air that will continue to resonate through the rest of the book: the fact that imaginative and dynamic thinking about the ether inevitably folded back on itself, to become a reflection on reflection itself. The ether is the ultimate aerial matter of thought.

Where Part One considers aspects of the motive or operative air, the air put to various kinds of physical and mental work, Part Two takes account of the ways in which, in the later part of the nineteenth century, the air became the arena of what Tyndall called an 'aetherial commotion'. The end of the nineteenth century saw the beginnings of an extraordinary change in the actual and imagined condition of the air. Increasingly, the traditional, poetic conception of the air as a kind of vacancy, the privileged bearer of the idea of the open itself, had to contend with a sense of the crowding of the air, which became liable to congestion, contamination and exhaustion. 'The atmosphere', a term that had provided a name for the open milieu of human actions and relations, suddenly became the name for that which broke in from the outside. 'Atmospherics' is the name for an interference, the air's interference with itself. The three chapters in this section explore different forms of this interference. The first considers the growing interest in haziness from the late nineteenth century onwards, and, construing haze as a kind of visual noise or dazzle, attempts to make out a general affinity between artisic modernism and the nebular. The second chapter moves from visual to acoustic space. It offers a history of radio not through its register of positive achievements, but through its struggle against the negative principle of atmospheric noise that plagued it from the beginning. We see that noise moving from the outside, as diabolical distraction that must at all costs be cleaned out from the radio signal, to the inside, first of all in the fact the human beings themselves are more and more the source of atmospheric interference, and secondly in the fact that atmospherics came into the foreground, as the focus of conscious attention – in the theremin, popularly understood as the making of music from static, in the artistic interest in the aurora borealis and other low-frequency phenomena in the atmosphere, and in applications like radio astronomy, in which accidental noise is promoted to the condition of signal. The final chapter in this phase of the book considers a more menacing form of air-pollution. Where in the past almost all funerary arrangements had understood the fact of death as a separation of earth from air, with the mortal body being returned to the dust from which it came, and the spiritual component allowed to return to its naturally elevated state, the growing interest in practices of cremation through the nineteenth and twentieth centuries had the effect of 'mortifying' the air, which became the repository and resting place of the dead, body and soul. At the same time, developments in military technology, especially the threat of poison gas and aerial bombardment, made more and more for 'an air that kills'.

Part Three offers analyses of two distinctly contemporary forms of our distinctive affinity with air and aerial process. The first is in the climate of the climactic furnished by the imagination of explosion, that most energetic form of the air's externalising or diffusive action. This chapter offers an account of the evolution of an intimate implication with explosive possibility that is a striking and uninvestigated part of the modern aerial imagination. The final chapter offers a discussion of the ideal of lightness as it is found in the delight in consuming levity, in the long dream of effervescence, given a new salience in the era of pop and an eminently bearable lightness of being.

One of the consequences of taking the air as a model for thinking about the history of the human relation to air itself is that it proves as difficult to make clean cuts between past and present, ancient and modern, as it is to make portion and partition of the air itself. If we are to understand social and cultural history as a kind of weather, and to understand the new understanding of the air as a kind of climate change, then it may be productive to bring to bear on this very process a kind of meteorological perspective, which is attuned to lingerings, minglings, perturbations and precipitations. A story told by the British politician Roy Hattersley about a ceremonial dinner in Pakistan he attended during the 1980s makes clear the capacity of the air to entertain different temporalities of belief:

> The head of Pakistan International Airlines, a former air chief marshal, was among the guests, and I could not resist the opportunity to ask him a contentious question. Earlier in the day I had been told – whether correctly or not I cannot say – that, according to the Qur'an, the sky is a blue carpet, held over Earth by Allah, and the stars His light which nothing could obscure.
>
> I thought that the idea was much more attractive than the explanation that I had been taught in schoolboy physics, but I could see that it might raise problems for devout believers. So, as the dinner progressed, I raised the subject of pious pilots. How did they deal with the idea of the carpet in the sky? There was, the airman gravely replied, no difficulty. 'They believe one thing up there and another down here.'[46]

But if there is no once and for all progression in the matter of the air, there is a regular fluctuation. Repeatedly, the air will be conceived in terms of an antinomy or argument between openness and limit. The many forms in which the air has been finitized have retroactively produced dreams the of absolute openness of the air, of the air as the absolute *apeiron*, the undetermined itself, and therefore as the material form of our own embattled dream of the aperture, the 'next degree to nothing' that we want the air to be, in order to continue to take ourselves for it.

PART ONE

Elaborations

two

A Very Beautiful Pneumatic Machinery

Just over two centuries ago, just as the machine began to impinge most closely upon human consciousness, and began to twin with it, machinery went mad. Or, what may come to the same thing, the mad became mechanical. They began to dream of machines, complicated dreams of sinister, far-reaching apparatuses. The mad machinery of these dreams is not a machinery out of control – not like the juridical machine of Kafka's *In the Penal Settlement*, which ends by systematically disassembling itself. It is a machinery that is mad in its functioning, its madness the madness of its very reliability, its repeatability. For the essential madness of the machine lies in its apparent rationality, the offer it holds out to rationality to find its like in mechanical process. For a sane man to believe seriously that he has become a machine is to become mad. A hundred years after what looks like its first appearance, Victor Tausk would give this machine a name, suggesting that all the machineries of the mad are varieties of or modular components of the mysterious, polymorphous apparatus he calls the 'influencing machine'. The purpose of this machine, which can be operated by persecuting entities over large distances, is to exert control over the patient's thoughts and feelings. Tausk explains that the influencing machine typically makes the patient see pictures (and when it does this, it takes the form of a magic lantern or cinematograph); it 'produces, as well as removes, thoughts and feelings by means of waves or rays or mysterious forces'; it produces motor phenomena in the body, like erections and emissions, by means of 'air-currents, electricity, magnetism, or x-rays'; it creates indescribable bodily sensations, 'that in part are sensed as electrical, magnetic, or due to air-currents'; and brings about other bodily symptoms such as abscesses and skin-eruptions.[1] But the most important function of the machine is its systematic dealing out of delusion, including, we must suppose, the master-delusion of the influencing machine itself. For the maddest thing about this mad contraption is that it is its own product, that its purpose seems to be to *manufacture, maintain and ramify itself.* Tausk makes it clear that the influencing machine always has a history, its workings becoming ever more complex and elaborate as the patient's psychosis becomes more deeply rooted. This is because the machine is also the patient's strongest defence against his madness, in providing a model and an explanation for how his madness works. The maddest thing about this machine is the knowledge it seems to provide of the workings of the madness, which is to say, of itself. Nowhere are the melancholy-mechanical mad madder than in the image they have and hold on to of their madness.

Psychoanalysis and the varieties of corporealized psychiatry similarly propose more or less mechanical systems as correlatives for the act of thinking. In fact, the influencing machine can be an exquisite parody of the psychoanalytic process itself, especially its tendency to convert the spiritual entities of previous eras (bad spirits and possessing demons) into impersonal forces and agencies – id, ego, superego, repression, cathexis, libido. There is thus a strange concurrence between illness and remedy in the case of mechanical delusions, or delusive machines. Tausk reports a number of cases in which patients felt flows and currents, including that of a man who 'felt electrical currents streaming through him, which entered the earth through his legs; he produced the current within himself, declaring with pride that that was his power!' But Tausk matches this conception in his explication of the processes behind the formation of the influencing machine, which requires us to accept the assumption 'that the libido flows throw the entire body, perhaps like a substance (Freud's view), and that the integration of the organism is effected by a libido tonus, the oscillations of which correspond to the oscillations of psychic narcissism and object libido'.[2] This flow of libido is far from merely metaphorical, since it is capable of bringing about transient swelling of organs, as a result of 'an overflow of secretion resulting from libidinal charging of organs'.[3] Both madness and its analysis are explications of the workings of machinery: of the machinery of delusion in the case of the physician and of the delusive machine of the patient, by which latter is meant the machine produced by delusion and the machine that deals out delusions, especially the delusion of itself.

Though he cannot resist the tendency to treat this machine as though it had an existence independent of its victim – indeed, he has a tendency to speak of it as if there were in fact only one machine – Tausk does not quite permit himself to take this machine seriously. It is after all a delusion. Since the machine is a cod, or counterfeit machine, since it cannot really work as it is supposed to work, Tausk does not feel it necessary to take its surface workings very seriously, to enquire into what kind of machine it may actually be. What matters most of all is to get behind the machine-as-symptom in order to understand the psychic machinery that has produced it. The machine is in the final instance mysterious, for a complete account of its workings cannot be given. Tausk proposes instead a genuine reading of the psychological mechanisms that can by contrast be exhibited and explicated in their totality.

Later explicators of the influencing machine have been more attentive to its details, and in particular to the technological forces and forms that it seems to incorporate. Readers of the elaborate delusions of Daniel Paul Schreber in particular have increasingly drawn attention to the importance of technologies of communication – telegraphs, telephones, radio – in his version of the influencing machine. The tendency has been to read all influencing machines as anticipations of the paranoia associated with such distinctively modern technologies of communication. I want to focus attention on a feature of the physical workings of the influencing machine that has not been subject to so much attention, namely the ways in which they are held to work over space and, more specifically, how they imagine the mechanical manipulation of what fills that space – whether air, ether or some other insubstantial matter. Conjugating a number of systematic accounts of systematic delusions, those in particular of James Tilly Matthews, Friedrich Krauss and John Perceval, I will try to construe the workings of the influencing machine as a pneumatics, or a fluid mechanics.

Gaz-plucking

In the autumn of 1809 a writ for habeas corpus was delivered to Bethlem Hospital demanding the release of an inmate who had been confined in the hospital for 13 years. This led to a hearing before the King's Bench in which affidavits from the family of the confined lunatic were considered against those furnished by Bethlem Hospital itself. Chief among the affidavits on the side of the confined man was an account by two doctors, one George Birkbeck and his associate Henry Clutterbuck, of their visits to the prisoner over the course of some months, leading to their conclusion that he exhibited no traces of insanity and should be released. Against them, the governors of the hospital brought forward various witnesses as to the confined man's dangerous insanity, which extended to making threats against the king and his ministers. Most tellingly of all, they quoted a letter of 7 September 1809 from no less a person than Lord Liverpool, the home secretary, recommending that the prisoner continue to be entertained at the public expense in Bethlem Hospital. The case for release was rejected.

It was not usual for the home secretary to be applied to in such a case. It seems likely that it was not in his office but in his person that Lord Liverpool was addressed. For this particular lunatic had first been brought to Bedlam following an outburst he had made on 30 December 1796 from the public gallery of the House of Commons during a speech made by Robert Banks Jenkinson, Lord Liverpool himself, defending the government against the charge that it had deliberately obstructed efforts to avert the costly and damaging war against France. Lord Liverpool's speech had been interrupted by a cry of 'Traitor!' from the gallery and a man had been hustled away by stewards. During his examination by the Bow Street magistrates, it emerged that the prisoner's name was James Tilly Matthews, who had a wife and young son. Matthews had been a tea merchant, but claimed to have been in the service of the government since just before the outbreak of war against France, acting as an emissary conducting secret peace talks with the revolutionary government. Now the government had decided against peace, seeing their opportunity of crushing decisively the political threat from across the Channel and halting the drift of revolution. Matthews's sense of betrayal had led to his public denunciation of Lord Liverpool, the minister whom he most particularly blamed for his predicament. The fierce threats that Matthews uttered against Lord Liverpool and other members of the government, combined with his obviously excitable condition, left the Bow Street magistrates with no choice but to request the governors of Bethlem Hospital to take custody of him.

One can perhaps understand why Lord Liverpool might have been disinclined to approve the release of Matthews even thirteen years after this event. The threat posed by somebody like Matthews seemed real. One of Matthews's fellow detainees in the Bethlem Hospital was James Hadfield, who had fired a pistol at George III in the Drury Lane Theatre on 15 May 1800 and, following the rapid institution of a law to permit the detention of the criminally insane, had also been committed to Bedlam. On 12 May 1812, a couple of years after Matthews's appeal was turned down, a man armed with a small pistol killed the prime minister, Spencer Perceval, in the lobby of the House of Commons. Twenty years later the son of the assassinated prime minister, who had been nine years old at the time of the murder, would himself be incarcerated in a lunatic asylum. As far as Lord Liverpool was concerned, insurrection was still, twenty years after the French Revolution, in the air. In this, he was in fact in close accord with James Tilly Matthews, especially as regards the air.

The story of what led to the outburst as a result of which James Tilly Matthews was confined in Bethlem Hospital, and why Lord Liverpool should have continued to take an interest in him, has been told a number of times, most recently and in the most fluent and illuminating detail by Mike Jay.[4] The most important thing about the 1810 habeas corpus hearing was the effect it had on John Haslam, the resident apothecary and senior medical officer of the Bethlem Hospital, spurring him into publishing an elaborate defence of his view that Matthews was completely insane and unfit to be released. Stung by the challenge to his professional judgement, Haslam was determined to show that Matthews was not only mad, but also epically and self-evidently so. The way to do this was, as he put it, 'to develop the peculiar opinions of Mr Matthews, and leave the reader to exercise his own judgement concerning them'.[5] His *Illustrations of Madness* accordingly gives Matthews's delusions the fullest possible airing, often using his own words, very likely taken down by Haslam himself, and giving a richness of detail unprecedented in the literature of madness. The result is a kind of accidental phenomenology, an insider account of Matthews's mad-machine and machine-madness. This is apparent even in the title of Haslam's book, which seems to slide from professional self-assurance into Matthews's idiosyncratic style, as though Haslam's and Matthews's voices were performing an awkward duet. The full title is: *Illustrations of Madness: Exhibiting A Singular Case of Insanity, and a No Less Remarkable Difference in Medical Opinion: Developing the Nature of Assailment, and the Manner of Working Events; With a Description of the Tortures Experienced by Bomb-Bursting, Lobster-Cracking, and Lengthening the Brain. Embellished With a Curious Plate.*

Haslam explained that Matthews believed himself to be subject to violent and continuous persecution by a gang of 'villains profoundly skilled in Pneumatic Chemistry', who operated from a basement not far from the hospital, and assailed him by means of an apparatus called an 'air-loom'. Haslam gives Matthews's roster of the seven members of the gang: 'Bill the King', their mysterious leader; 'Jack the Schoolmaster', a shorthand writer who records all the gang's doings; Sir Archy, a foul-mouthed blackguard, who is possibly a woman dressed as a man; the 'Middle Man', the most skilful operator of the instrument; 'Augusta', a 36-year-old woman, who is frequently out and about; 'Charlotte', apparently French, and kept a prisoner by the gang; and finally one who has no name but the 'Glove Woman', who keeps her arms covered because of the itch, operates the machine with skill, but has never been known to speak. Another document of Matthews, not quoted by Haslam, but included in Roy Porter's edition of *Illustrations of Madness*, suggests that there may be more even than this, for he speaks of '[t]he dreadful gang of 13 or 14 Monster Men & Women who are so making their Efforts on Me' (*Illustrations*, lxiii).

The most extraordinary feature of Haslam's report is the drawing, by Matthews himself, which he provides of the air-loom, along with Matthews's own detailed key to its elements. The machine seems to

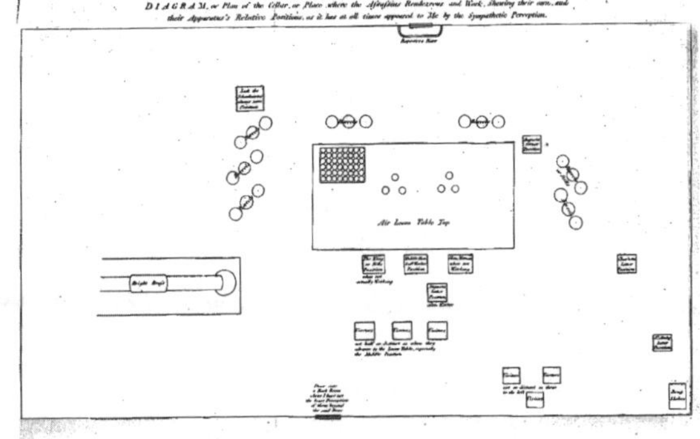

James Tilly Matthews's drawing of the air-loom gang at work, from James Haslam, *Illustrations of Madness* (London, 1810).

have worked in something like the following fashion. Noxious vapours contained in barrels are directed through pipes into the principal body of the apparatus, which is contained invisibly within an enormous desk-like structure (one of a number of features in the machine that suggests the parallels between the machine and the psychic bureaucracy that a later and still much more celebrated paranoiac, Daniel Paul Schreber, would call an *Aufschreibesytem*, a 'writing-down-system').[6] As a result of some mysterious process of distillation occurring inside the machine, a magnetic-mesmeric fluid is produced, and perhaps stored in the battery-like 'cluster of upright open tubes or cylinders, and, by the assassins, termed their *musical glasses*' (*Illustrations*, 45). These objects may also be meant to be the magnets to which Matthews refers, and which do not seem otherwise to be visible in the illustration. It is drawn off through the tubes emerging from the apparatus and transmitted by a 'windmill kind of sails' (*Illustrations*, 45). The substance itself is described as

> The warp of magnetic-fluid, reaching between the person impregnated with such fluid, and the air-loom magnets to which it is prepared; which being a multiplicity of fine wires of fluid, forms the sympathy, streams of attraction, repulsion, &c. as putting the different poles of the common magnet to objects operates; and by which sympathetic warp the assailed object is affected at pleasure. (*Illustrations*, 48)

Edmund Cartwright had invented a power-loom that could be operated by steam or water in 1785. Matthews's machine still appears to be largely hand-operated, and its operation seems to require considerable finesse on the part of the gang. Matthews draws attention to the levers, 'by the management of which the assailed is wrenched, stagnated, and the sudden-death efforts made upon him' (*Illustrations*, 43). He also refers to '[t]hings, apparently pedals, worked by the feet of the pneumaticians' (*Illustrations*, 45) and '[s]omething like piano-forte keys, which open the tube valves within the air-loom, to spread or feed the warp of magnetic fluid' (*Illustrations*, 42–3), though these last do not appear in Matthews's illustration.

The effects of the fluid on its victim are exotic and fearsome. They include: '*Fluid-locking*', a constriction of the tongue which prevents speech; '*Cutting soul from sense*', in which the heart and the intellect are separated by a veil of magnetic fluid (*Illustrations*, 30); '*Stone-making*', the formation of calculus in the bladder (Matthews may perhaps have heard of experiments in the early days of pneumatic therapy that gallstones might be dissolved by the breathing in of certain gases, or the drinking of waters impregnated with carbon dioxide following Priestley's method of manufacture);[7] '*Thigh-talking*', in which the victim becomes convinced that 'his organ of hearing, with all its sensibility', is lodged on the external portion of the thigh, to which utterances are magnetically directed.

Matthews is also subject to the violently kinetic effects of pneumatics. The most fiendish torture of all (the one in which the operator in the illustration is in fact engaged) is called '*Sudden-death-squeezing*' or '*Lobster-cracking*', described as 'an external pressure of the magnetic atmosphere surrounding the person assailed, so as to stagnate his circulation, impede his vital motions, and produce instant death' (*Illustrations*, 32). There is also '*Laugh-making*', which 'consists in forcing the magnetic fluid, rarified and subtilized, on the vitals [*vital touching*] so that the muscles of the face become screwed into a laugh or grin' (*Illustrations*, 35). Not only is he subject to impregnation with the magnetic fluid, the operators make him a pharmaceutical factory for the production of more fluid:

> *Gaz-plucking* is the extraction of magnetic fluid from a person assailed, such fluid having been rarified and sublimed by its continuance in the stomach and intestines. This gaz is in great request, and considered as the most valuable for the infernal purposes of these wretches. They contrive, in a very dexterous manner, to extract it from the anus of the person assailed, by the suction of the air-loom. This process is performed in a very gradual way, bubble by bubble. (*Illustrations*, 37–8)

We are left to imagine for ourselves the exquisite tortures encompassed in the terms '*foot-curving, lethargy-making, spark-exploding, knee-nailing, burning out, eye-screwing, sight-stopping, roof-stringing, vital-tearing* [and] *fibre-ripping*' (*Illustrations*, 38).

Victor Tausk decided that the influencing machine was a dissimulated image of the patient's own body, and more specifically his genitals. The growing complexity of the machine was determined by the alternation of inhibition and gratification that it was designed to mediate.[8] We may suppose with Tausk that there is always a close relation, of resemblance, dependence, disavowal or antagonism, between a machine and the body. Before there were machines, there were mechanical theories of the body. And this machinery was not only kinetic, but also pneumatic and chemical. If bodies are like machines, then perhaps all machines are intelligible as bodies, and the efforts to understand the body, especially those that understand it in mechanical terms, have made it a fertile generator of imaginary machines and delirious devices. The importance of the entry of machinery into madness is that it seems to dramatize the moment at which the increasing familiarity of machines reactivated a tradition of fabular machines. Visionary and phoney contraptions, like the hoax automata interestingly and confusingly made by builders of real automata, the speaking machines and the Chess-Playing Turk, began to propagate alongside real ones.

But, for all the horrifyingly corporeal tortures it inflicts, the air-loom is much more intelligible as a reflection on Matthews's mind, or on the processes of thought, though indeed the result is a dramatically corporealised thinking. Indeed, we might almost see the air-loom principally as a machine for converting things mental into physical form, and usually in a form involving the effects of air on the body. So '*Kiteing*' is explained as the lifting into the brain, like a kite on a string, of 'some particular idea, which floats and undulates in the intellect for hours together . . . [and] keeps waving in his mind, and fixes his attention to the exclusion of other thoughts' (*Illustrations*, 31–2). Loss of concentration is ascribed to the effects of '*Bladder-filling*', which is 'filling the nerves of the neck with gaz, and by continued distension, effecting a partial dislocation of the brain. This frequently repeated, produces weakness of intellect' (*Illustrations*, 36). Even Matthews's efforts to articulate his own consciousness of the machine are subject to its workings, as in the operation known as '*Poking, or pushing up the quicksilver*':

> While in the act, as he supposes, of venting the burst of indignation, they contrive to push a seeming thread of fluid through his back diagonally in the direction of his vitals. Its operation is instantaneous, and the push appears to elevate the fluid about half an inch. This magic touch disarms the expression of his resentment, and leaves him an impotent prey to the malignity of their scorn and ridicule. (*Illustrations*, 35–6)

The clearest example of the action of corporealization is Matthews's Haslamized account of '*Lengthening the brain*', in which the brain and its mental contents are collapsed together:

> As the cylindrical mirror lengthens the countenance of the person who views himself in such glass, so the assailants have a method by which they contrive to elongate the brain. The effect produced by this process is a distortion of any idea in the mind, whereby that which had been considered as most serious becomes an object of ridicule. All thoughts are made to assume a grotesque interpretation; and the person assailed is surprised that his fixed and solemn opinions should take a form which compels him to distrust their identity and forces him to laugh at the most important subjects. (*Illustrations*, 33–4)

It is a surprising fact that all thought is unconscious, in the sense that we are no more aware of the physical processes that give rise to our thinking than we are of the circulation of our blood, or the secretions from our kidneys. This kind of corporealization of thought seems to give Matthews access to the mechanical processes of his own thought, making it possible for him to see and feel directly his thought in operation. The air-loom is an allegory of the process of thought, perhaps an allegory of the process of imaging, of allegorizing thought, of making matter of it.

Mind, Matter, Metaphor

Making the processes of thought accessible to the mind seems to involve a move from the dominion of the ear to that of the eye. In his *Observations on Madness and Melancholy*, which appeared in the year before the Matthews affair blew up, John Haslam had stressed that

> [o]f the organs of sense, which become affected in those labouring under insanity, the ear, more particularly suffers. I scarcely recollect an instance of a lunatic becoming blind, but numbers more deaf. It is also certain that in these persons, more delusion is conveyed through the ear than the eye, or any of the other senses.[9]

Interestingly, he gives hints of the ways in which other patients suffering from aural persecution resorted to the idea of various kinds of apparatus in order to process and concretize the otherwise terrifyingly inchoate experience of their auditory hallucinations:

> in consequence of some affection of the ear, the insane sometimes insist that malicious agents contrive to blow streams of infected air into this organ: others have conceived, by means of what they call hearkening wires and whiz-pipes, that various obscenities and blasphemies are forced into their minds.[10]

This is the first of a number of occasions in which the idea of pneumatic breaking and entering through the ear will make an appearance in this discussion. Perhaps those 'hearkening wires and whiz-pipes' are incipient forms of the influencing machine that has come to much fuller development in Matthews's delusion.

But there is another feature of the air-loom, apparent not only in Matthews's but also in other such systematic delusions, that is rarely discussed, though all readers from Haslam onwards are surely aware of it, namely the fact that the whole apparatus, the whole conception of the air-loom, is so richly and deliciously risible. Matthews must surely have been aware of the arguments of magnetizers and mesmerists that the vital fluid that they were manipulating was supposed to derive its power from its ethereal fineness. And yet this superfine stuff is supposed to be distilled from the scavenged stinks contained in the barrels that decant into the machine, which are much more the kind of thing one would expect to find in a witch's cauldron than in the workshop of a pneumatic technician:

> Seminal fluid, male and female – Effluvia of copper – ditto of sulphur – the vapours of vitriol and aqua fortis – ditto of nightshade and hellebore – effluvia of dogs – stinking human breath – putrid effluvia – ditto of mortification and of the plague – stench of the sesspool – gaz from the anus of the horse – human gaz – gaz of the horse's greasy heels – Egyptian snuff (this is a dusty vapour, extremely nauseous, but its composition has not been hitherto ascertained) – vapour and effluvia of arsenic – poison of toad – otto of roses and of carnation. (Illustrations, 28–9)

Of course, the comedy may in large part be an effect of Haslam's narration, the purpose of which is to make the desperate looniness of Matthews's notions as palpable as possible. But the effort to read against the grain of Haslam's heavy irony, in order to retrieve Matthews's uncontaminated voice, has led to an unnecessary purging of what may be Matthews's own sense of the farcical nature of the whole affair, which is evidenced, for example, in his contemptuous rendering of the dialects of his persecutors. Though Matthews stresses the fact that the machine is built on the basis of the most fearsomely up-to-date science, what strikes us most markedly in its representation is the grotesque collage of the ancient and the modern, the efficient and the laborious, the cerebral and the corporeal. A machine is a reproduction and often also a grotesque parody of living process (hence the offensiveness of representing living process as 'mere' mechanicity). Matthews seems to give us a parody of the very idea of a machine, which seems to draw attention to its own status as sorry fantasy. It is almost as if Matthews were giving us the very image of an obviously spoof machine, the purpose of which is to demonstrate its own implausibility.

The comedy of Matthews's machine derives from its corporeality, from the Bergsonian recoil of the *élan vital* from the idea of human thought being subjugated by a series of gimcrack contrivances like this one. Curiously, though, there seems to be a kind of meeting of minds between Matthews and Haslam on this score. Both of them have a tendency to view the operations of the mind in bizarrely material terms, and they seem to agree that the disordering of sense and idea characteristic of madness is a result of strictly physical causes. As one might perhaps expect from the superintendent of an asylum, Haslam does not have a very high opinion of the powers of the human mind. As he had observed in his earlier *Observations of Madness and Melancholy*:

> As far as I have observed respecting the human mind, (and I speak with great hesitation and diffidence,) it does not possess, all those powers and faculties with which the pride of man has thought proper to invest it. By our senses, we are enabled to become acquainted

> with objects, and we are capable of recollecting them in a great or less degree; the rest, appears to be merely a contrivance of language. (OMM, 9)

The limitation of the mind's powers is nowhere more painfully apparent than when it tries to reflect upon its own constitution or operations:

> If mind, were actually capable of the operations attributed to it, and possessed these powers, it would necessarily have been able to create a language expressive of these powers and operations. But the fact is otherwise. The language, which characterizes mind and its operations, has been borrowed from external objects; for mind has no language peculiar to itself. (OMM, 9)

'[W]e live in a world of metaphor', Haslam confirms in a later footnote (OMM, 34). It is for this reason that he has so little faith in reasoning with the mad, or dealing with them on their own terms, for this would mean giving credence to the delusions from which they need to be delivered. This is particularly the case with 'methodical madness' like that of Matthews (though Matthews does not seem to be mentioned or referred to in this earlier text), since '[i]n proportion as insanity has assumed a systematic character, it become [*sic*] more difficult of cure' (OMM, 269). Haslam does not believe there can be such a thing as a disease of the mind, since the mind is incorporeal. However, a 'disease of ideas', caused by a disordering of the material in which the mind is embedded (and from which it may be required to derive its ideas of itself), is conceivable, though also deeply intractable.

The case histories in *Observations of Madness and Melancholy* show that, even before his account of Matthews, Haslam had an interest in the systematic forms of delusion, though he tends to present these for the amusement of the reader. The most developed account of this kind to be found in the book is that of a patient who suffered from delusions of mind control that have similarities with those of Matthews, though machinery has somewhat less of a role in his delusions – or at least in Haslam's account of them. The patient in question used to stop his ears with cotton wool and place a saucepan over his head while he slept, in order to prevent the entrance of what he called 'sprites'. These were formed, evidently by spontaneous generation, from the semen of various 'occult agents' that they preserved in rum or brandy, and then introduced into the ears of their victims. Haslam gives his patient's own account of their workings:

> As the semen in the natural commerce with the woman, produces a child, so having its vitality conserved by the spirit, it becomes capable of forming a sprite; a term, obviously derived from the spirit in which it had been infused. The ear is the most convenient nidus for hatching these vital particles of the semen. The effects produced on the individual, during the incubation of these seminal germs, are very disagreeable; they cause the blood to mount into the head, and produce considerable giddiness and confusion of thought. In a short time, they acquire the size of a pin's head; and then they perforate the drum of the ear, which enables them to traverse the interior of the brain, and become acquainted with the hidden secrets of the person's mind. During the time they are thus educated, they enlarge according to the natural laws of growth; they then take wing, and become invisible beings, and from the strong ties of natural affection, assisted by the principle of attraction, they revert to the parent who afforded the semen, and communicate to him their surreptitious observations and intellectual gleanings. In this manner, I have been defrauded of discoveries which would have entitled me to opulence and distinction, and have lived to see others reap honours and emoluments, for speculations which were the genuine offsprings of my own brain. (OMM, 73–5)

Haslam seems to want us to appreciate the specifically material character of this delusion, which represents mental processes in terms of 'external objects'. The sufferer from this delusion, like Matthews himself, seems to have been invaded, not by sprites, but by metaphors, metaphors that create the delusion of a mind made of nothing but matter.

And yet Haslam himself is convinced that madness is a corporeal condition. Accounts of patients' symptoms are accompanied by descriptions of the results of brain dissections, which repeatedly report the symptoms of various kinds of pressure or swelling in the brain. The physical causes suggested by Haslam include intoxication, blows to the head, mercury poisoning and 'cutaneous eruptions repelled, and the suppression of periodical or occasional discharges and secretions' (OMM, 209). Haslam's earlier *Observations on Insanity* of 1798, of which *Observations on Madness and Melancholy* is a revision and development, regularly notes, accompanying the enlargement of the brain by fluid, the presence of air in the veins: 'there was a striking appearance of air in the veins'; 'the veins seemed to contain air'; 'the vessels of the pia mater were turgid, and its veins contained a quantity of air'.[11] It seems as though, from Haslam's point of view, the delusions of invasion by tainted or magical air may be the effect of an actual impregnation of the brain and its blood vessels by air or other foreign element – and thus the delusion may accidentally constitute an accurate self-representation of the physical deficit of the organ of thought.

Haslam is not alone in his interest in the cognitive effects of body gas. His views echo, and may even have helped form, those of Thomas Beddoes, who, after his long involvement with pneumatic medicine, had turned his attention to the problem of nervous diseases in his *Hygëia* (1803), a collection of essays on the social basis of health and disease. Beddoes too is interested in identifying what may be the physical causes of mental illness, but notes that 'here morbid anatomy is at present of little or no use in pointing out the correspondence between the obvious and hidden workings of the machine'.[12] But he does see one area of hopeful advance, noting that '[c]onsiderable discoveries have however been made in pneumatology' (*Hygëia*, III, 15). During his experiments at the Pneumatic Institution which he founded in Bristol in 1798. Beddoes had for some time been persuaded that over-oxygenation might be to blame for consumption – he even thought that 'a phthisical patient would take a longer time than another person in being drowned'[13] – and so had treated patients with carbon dioxide, hoping that this would help lower oxygen levels in their blood and lungs. Beddoes held until the end of his life to the value of having sufferers from tuberculosis sleep in cowsheds, to reap the benefit of the oxygen-depleted air.[14] He even ascribed the increase in consumption in Britain to the growing salubriousness of the national atmosphere:

> The inhabitants of this country, almost without exception, breathe a freer and a purer air than their ancestors . . . You see then that the subjects of our Edwards, and our Henrys and of good Queen Bess may have found, in being free from so formidable a disease than their delicate and airy posterity, some compensation for the confined air and filth in which they passed their existence.[15]

The air was the most important mediator between body and mind for Beddoes, and therefore the most important of the regulators of the two leading principles of his Brunonian medical theory, excitability and lethargy, or the sthenic and the asthenic.[16] Beddoes thought that nervous diseases such as epilepsy were also caused by excess of oxygen, remarking of the narrative of a German sufferer from epilepsy, that 'The facts are at least curious in a view to pneumatology' (*Hygëia*, III, 118). He drew attention to the airy associations of the sensation known as the 'aura', which heralded an epileptic seizure, remarking that '[t]hese wandering sensations, as of a breeze passing along parts of the body, may have procured to nervous affections, the name of *vapours*' (*Hygëia*, III, 198). He seems to have thought it possible to regulate mental states by preventing over-stimulation through oxygen carried in the blood:

> . . . compression of arteries that carry blood up to the neck, has been found to put a temporary stop to nervous fits in a case where palpitation of the heart, headache, coldness of the feet, occasional shivering followed by extreme heats, particularly about the face and head, locked jaw, convulsions of the muscles of the neck and body, difficulty of breathing, stupor and delirium succeeded each other. (*Hygëia*, III, 119n)

He thought that the thickened arteries of epileptics were the effect of their illness: 'Frequent throbbings, flushings, and heats about the head, the tokens of too great action in its arteries, must, from all analogy, be regarded as capable of thickening the bones, and producing an alteration in the structure of all the parts' (*Hygëia*, III, 117). He advised sleeping sitting up, to reduce blood-flow to the head (*Hygëia*, III, 117), along with various kinds of mechanical routine, like speaking slowly and deliberately, to forestall 'the formidable *hurry of ideas*' characteristic of the onset of epilepsy (*Hygëia*, III, 118). Drinkers, by contrast, suffered from dangerously low levels of oxygen, which Beddoes indicates by the case of 'the inflammable woman of Coventry . . . [who] seems to have reduced herself by dram-drinking to such a state as to be capable of being set on fire by a spark'.[17] Beddoes has an intriguing theory to account for this: '[O]ne is justified by all the known facts relative to combustion, in supposing, that where the substance of the body was so eager to combine with external oxygene, there must have been an internal deficiency of this element.'[18]

Although Beddoes writes a good deal of sense regarding the desirability of hygiene, exercise and a moderate diet, he also has some obsessively mechanistic ideas regarding the physical causes of nervous diseases, ideas that revert often to what he calls the 'pneumatological'. There was, he thought, a close connection between consumption and nervous complaints. 'Such are the vicissitudes of the atmosphere of these islands, and such the tendency of our habits', he wrote, 'that constitutional weakness will very frequently be visited upon the lungs; and more frequently still, will constitutional weakness shew itself by tremors, startings, wakefulness, fatiguing sleep and by other indications' (*Hygëia*, III, 145–6). Exposure to fresh air and the avoidance of draughts were principles to be followed with almost superstitious reverence; he warns that 'I have known slight passing stupors converted into epilepsy by the shower-bath' (*Hygëia*, III, 146–7). Though the younger Beddoes had been associated with radical ideas, his later writings show a crusty suspicion of the over-stimulations of modern life, especially the 'hubbub in the brain' caused by too much light reading:

> What wonder then that we should hear complaints against the age as wanting energy of feeling and compass of mind? What

> wonder that while idea reels against idea, we should so often experience an analogous unsteadiness of footing? . . . [Y]ou will understand how possible it is, that a variety of prevalent indispositions, as *fluor albus*, tendency to miscarriage, and even a dropsy of the ovarium, may be caught from the furniture of a circulating library. (*Hygëia*, III, 163–4, 165)

The best thing for young people would be

> to lay the use of books pretty nearly aside, and during the season of the mind and body, to trust principally to the senses, directed by the living voice – that living voice, which, as Quintilian says, affords substantial nourishment, while books (it is certainly the case with much reading) do little perhaps but stimulate . . . The juvenile library! With submission, I must consider it as little better than a repository of poisons. (*Hygëia*, III, 166)

Beddoes had made some very large claims at the beginning of his career regarding the benefits of the control of the air:

> The more you reflect, the more you will be convinced that nothing would so much contribute to rescue the art of medicine from its present helpless condition, as the discovery of the means of regulating the constitution of the atmosphere. It would be no less desirable to have a convenient method of reducing the oxygene to 18 or 20 in 100, than of increasing it in any proportion. The influence of the air we breathe is as wide as the diffusion of the blood. The minutest portions of the organs of motion, sense, and thought must be affected by any considerable change in this fluid.[19]

He even looked forward to the enhancement of intellectual life through a kind of pneumatic engineering of respirable air, suggesting that 'perhaps there may be a mixture of azotic and oxygene airs more favourable to the intellectual faculties that [*sic*] that which is found in the atmosphere; and hence chemistry may be enabled to exalt the powers of future poets and philosophers.'[20] Erasmus Darwin's poem in praise of Beddoes's prescriptions of gaseous regimes and remedies involuntarily suggests the immoderate fantasy to which his pneumatological materialism could attain:

> Happy, thrice happy he, who at his will
> Can drink of Life's sweet cup his constant fill;
> Who, if excess of Oxygene create
> Symptoms, which lean Consumption indicate,
> A sure specific can procure with ease,
> Rich Cream and Butter from his herd of Trees:
> Or if he find excess of Hydrogene
> His body load with fat, his mind with spleen,
> True health and vigour to restore, can take
> From some regenerate Oak a savoury steak.[21]

Following the disappointing results of Beddoes's experiments with pneumatic therapeutics in the late 1790s, his essays on nervous diseases in *Hygëia* are much less confident. The hyperventilating, over-stimulated subjects of the period must endure domestic versions of the kind of hypochondriac, hysterical assailments suffered by Matthews:

> I am afraid that in this jarring and tumultuous world, the poor sensitive human plant will be utterly at a loss to find an asylum. Wherever he retires, the occasional causes of his paroxysms, be they epileptic, hysterical, *cephalalgic*, or anomalous, will pursue and hunt him out. Earthquakes, volcanic eruptions, hurricanes, revolutions, mobs, he may have the good fortune, all his life, to escape. But he will find the whole cultivated part of the earth infested with heats, frosts, storms, thunder-claps, door-claps, creaking wheels, jarring windows, squeaking pigs, cackling geese, crowing cocks, and unnumerable disorders besides in the apparatus of polished society. (*Hygëia*, III, 201)

Not surprisingly, perhaps, Matthews was much less suspicious of fresh air than Beddoes. Mike Jay usefully reminds us that sanitary conditions in Bethlem were indeed extremely poor, and draws attention to Matthews's own concern for fresh air and ventilation, when he came in 1811 to submit architectural plans for the projected rebuilding of the hospital. His plans emphasized the necessity for elevation, so that patients could feel the restorative benefit of the winds: Matthews proposed that the female patients should occupy the west wing 'as being open to the warmest and most salubrious winds, the men being better able to endure the bleaker air'.[22]

Chemical Revolution

The importance of the air both in Beddoes's pneumatology and in Matthews's delusions is that the air is traditionally the matter of thought. Hence the predominantly pneumatic or vaporous machinery of late medieval psychology. The pneumatic mechanics of the seventeenth century and the pneumatic chemistry of the eighteenth were a decisive epoch in what may be called the materialization of the air. As air passed over from the side of the subject to that of the object, the air changed from a principle of infinitude into a series of finite objects. Since air has the particularly intimate affinity it does with thought, this may be read as a reflexive doubling, a subduing of thought to and by itself.

This short-circuit is focused in the increasingly divergent fields of reference of the word 'pneumatic'. Among philosophers and theologians, the pneumatic referred to things of the spirit. In 1744 there was still a Chair in Ethics and Pneumatic Philosophy at the University of Edinburgh (David Hume put in for it, but was turned down), and the University of St Andrews instituted a Chair in Ethics and Pneumatics in 1747. But, from the seventeenth century onwards, the word had also increasingly been put to use in the most exciting new area of experimental science, the 'pneumatic chemistry' that was exploring the nature and behaviour of gases.

During the 1780s and 1790s pneumatic science had a progressive, not to say revolutionary air. While Matthews was beginning the long period of incarceration in Paris that would definitively evict him from his wits, Beddoes was at work at his Pneumatic Institution, which he

Joseph Priestley's apparatus for experimenting on gases, from his *Experiments and Observations on Different Kinds of Air* (London, 1775).

had established in Bristol in 1798. Like Joseph Priestley, who acquired a reputation as a political radical, and ended up being driven out of the country after an anti-French mob had burnt his home to the ground, Beddoes was a supporter of the French Revolution. Both saw links between the advance of the new gas chemistry and radical political reform. In 1790 Priestley had thundered that 'the English hierarchy, if there be anything unsound in its constitution, has equal reason to tremble before an air pump, or an electrical machine'.[23] After Beddoes left Oxford in 1793, depressed by the atmosphere of conformity there, his outspoken protests against Pitt's repressive measures earned him a reputation for seditious radicalism. He failed to gain the support of Joseph Banks, the president of the Royal Society, for his Pneumatic Institute; in a letter to Georgiana, Duchess of Devonshire, on 30 November 1794, Banks voiced his doubts 'concerning the propriety of his giving public countenance of any kind [to] a man who has openly avowd opinions utterly inimical in the extreme to the present arrangement of the Order of Society in this Country'.[24] Beddoes attempted to withdraw from politics between 1793 and 1795, but acknowledged, in a letter of December 1795, that 'my politics have been injurious to the airs'.[25] The association between gas and revolution was sealed by Edmund Burke, in the opening pages of his *Reflections on the Revolution in France* (1790):

> When I see the spirit of liberty in action, I see a strong principle at work; and this, for a while, is all I can possibly know of it. The wild *gas*, the fixed air, is plainly broke loose: but we ought to suspend our judgment until the first effervescence is a little subsided, till the liquor is cleared, and until we see something deeper than the agitation of a troubled and frothy surface.[26]

The italicization of that 'gas', followed by the more homely phrase 'fixed air', encodes the difference between French and English political chemistry. Boyle, Black, Priestley and Davy had used varieties of the word 'air' to distinguish from common air the elastic fluids they produced in their experiments. The word 'gas', which had been introduced by Van Helmont, suffered perhaps from the wildness of Van Helmont's mystical reputation, and certainly was associated with French rather than English usage. In 1787 four French chemists announced the result of eight months' intensive work to rationalize and modernize the nomenclature of chemistry, clearing away the picturesque but confusing names inherited from medieval alchemy.[27] It was translated into English by James St John the following year as the *Method of Chymical Nomenclature*. Among the most important of the changes proposed by the 'chemical revolution' was the use of the word 'gas' in place of the terminology of 'aeriform fluids', 'elastic fluids' and 'airs' used by earlier investigators, and commonest among British chemists. So 'fixed air', also known as 'mephitic gas' and 'solid air', became 'carbonic acid gas' (later carbon dioxide); 'inflammable air' became 'hydrogen'; 'marine air' became 'muriatic acid gas' (later hydrogen chloride, which becomes hydrochloric acid in solution); 'vitiated air' became 'azotic gas' (later nitrogen); and, most importantly, 'vital air', or 'dephlogisticated air', became 'oxygen'.[28]

The introduction of this last name disclosed another fault line between the French and the English understandings of gas, namely the phlogiston theory. Joseph Priestley, who had been the first to isolate oxygen from the air, by heating mercury calyx, held until late in his life to the phlogiston theory, which explained combustion as the giving off of a substance into it, rather than an extraction of a substance from it. In the opening words of his *Considerations on the Doctrine of Phlogiston and the Decomposition of Water* (1796), Priestley reminded his readers that 'There have been few, if any, revolutions in science so great, so sudden, and so general, as the prevalence of what is now usually termed *the new system of chemistry*, or that of the *Antiphlogistians*', which 'is often called that of the *French*'.[29]

It certainly seems to be the case that 'oxygen' became a kind of revolutionary code-word. And yet we should note that Burke's reference is not just to 'wild gas', a reference to the 'spiritus silvestris' that had been Van Helmont's name for carbon dioxide, but also to a more sedately native air, namely 'fixed air', the name given by Joseph Black to carbon dioxide when he discovered it in 1754. Burke's 'troubled and frothy surface' of contemporary events is in fact a reference to the effervescence of beer, which was a well-known source of what was not yet known as carbon dioxide. In fact, as its name might suggest, 'fixed air', so named because it seemed to be so much less reactive than ordinary air, was thought of as having preservative rather than explosive properties. It was noticed that hanging rotting meat over effervescing solutions seemed to slow or halt its decomposition, and the gas got a reputation for being useful in the treatment of what were called 'putrid fevers'.[30] The belief in the powers of fixed air persisted for some considerable time. A visitor to London in 1809 reported that 'Now we hear little of chalybeates, [iron-impregnated water] but fixed air is to conquer disease, and set death at defiance'.[31]

And yet, later, in *Reflections on the Revolution in France*, Burke gives us an alternative image of the wild air, in the course of a discussion of the folly of destroying all existing institutions in the interests of sweeping away injustice. For a reforming politician to do this is 'to consider his country as nothing but *carte blanche*, upon which he may scribble whatever he pleases'. The wise reformer will look 'for a

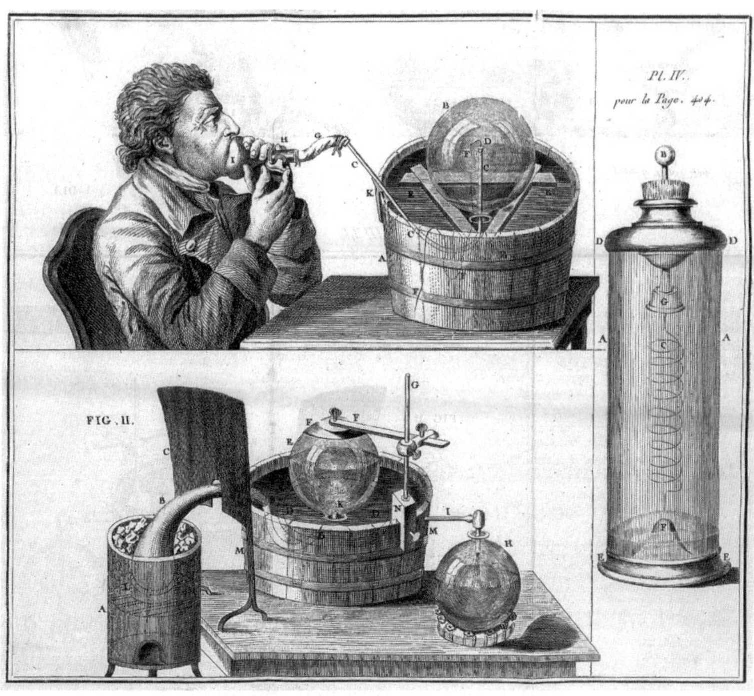

An apparatus for inhaling dephlogisticated air, from Jean Ingen-Housz, *Nouvelles expériences et observations sur divers objets de physique* (Paris, 1785).

power, what our workmen call a *purchase;* and if he finds that power, in politics as in mechanics, he cannot be at a loss to apply it.' Among the mechanical principles enumerated by Burke is that associated with the 'fixed air' mentioned in his opening pages:

> To destroy any power, growing wild from the rank productive force of the human mind, is almost tantamount, in the moral world, to the destruction of the apparently active properties of bodies in the material. It would be like the attempt to destroy (if it were in our competence to destroy) the expansive force of fixed air in nitre, or the power of steam, or of electricity, or of magnetism. These energies always existed in nature, and they were always discernible. They seemed, some of them unserviceable, some noxious, some no better than a sport to children; until

> contemplative ability, combining with practic skill, tamed their wild nature, subdued them to use, and rendered them at once the most powerful and the most tractable agents, in subservience to the great views and designs of men.[32]

The 'expansive force of fixed air in nitre' may refer to the force of explosion, since nitre is a name for saltpetre, the principal ingredient of gunpowder (though the gas it gives off is oxygen rather than carbon dioxide). Not surprisingly, explosion often came to mind as an image of the bursting out of revolutionary force:

> e'en now the vital air
> Of Liberty, condens'd awhile, is bursting
> (Force irresistible!) from its compressure
> To shatter the arch chemist in the explosion![33]

But there was another kind of revolution associated with the new pneumatic chemistry. The wildest of the 'airs' investigated by Beddoes, and probably also the one that was most injurious to his reputation, was not oxygen but nitrous oxide, which was first seriously studied by the young Humphry Davy in collaboration with Beddoes. As well as subjecting the gas to detailed chemical investigation, Davy also inhaled it regularly and in large quantities to see its effects on himself. Beddoes and Davy had made the acquaintance of Coleridge and Southey, who were among their experimental subjects.

The gas caused what Beddoes called 'high orgasm' of the muscles, in the form of quivering and tingling.[34] 'I felt a thrill in my teeth', recorded Southey.[35] (One might speculate enjoyably about the continuity between the vibratory modes signalled in the poetic word 'thrill' and the word 'buzz', which became common usage to signify excitement, specifically of intoxication, during the later twentieth century.) This was often expressed in an irresistible desire to giggle (hence its later name, 'laughing gas'). It intensified sight and hearing and gave a sense of delicious spaciousness and sublime exhilaration: a Mr Wedgwood said that 'I felt as if I were lighter than the atmosphere, and as if I was going to mount to the top of the room'.[36]

'The Whimsical Effects of Nitrous Oxide Gas', frontispiece to John Joseph Griffin, *Chemical Recreations: A Series of Amusing and Instructive Experiments*, 7th edn (Glasgow and London, 1834).

Beddoes recorded that, after administration of nitrous oxide, his wife 'could walk much better up Clifton Hill – has frequently seemed to be ascending like a balloon'.[37] There seemed to be few deleterious effects, except on ladies with a history of hysteria, itself, of course, conceived as a highly vaporous condition of the body, whom it sometimes sent off into fits.[38] One 'Miss N.' suffered for some weeks from '[l]aughing, crying, starting, painful acuteness and dullness of hearing, inordinate motion of almost every muscle, with loss of power in each, expectation of immediate death . . . She had no idea air could act so banefully'.[39] Its only effect on Coleridge, who was more accustomed perhaps to the hard stuff, was to cause him to stamp his feet on the floor uncontrollably.

From the beginning, the gas was interpreted in poetic or metaphysical terms. Southey was guarded in the account of his experiences he wrote to Beddoes and Davy, but was much less so in a letter he wrote to his brother: 'Oh, excellent air-bag! Tom, I am sure the air in heaven must be this wonder-working air of delight!'[40] Davy, who, after his first experiments in April 1799, quickly developed a taste for the gas, and could not see the silken air-bag in use without the craving for a whiff coming on, had himself enclosed in an air-tight inhalation-box (appropriately enough on 26 December), to inhale 20 quarts of nitrous oxide. The effects were spectacular:

> A thrilling extending from my chest to the extremities was almost immediately produced. I felt a sense of tangible extension highly pleasurable in every limb; my visible impressions were dazzling and apparently magnified, I heard distinctly every sound in the room and was perfectly aware of my situation. By degrees as the pleasurable sensations increased, I lost all connection with external things; trains of vivid visible images rapidly passed through my mind and were connected with words in such a manner, as to produce perceptions perfectly novel. I existed in a world of newly connected and newly modified ideas. I theorised; I imagined that I made discoveries . . . My emotions were enthusiastic and sublime; and for a minute I walked around the room perfectly regardless of what was said to me. As I recovered my former state of mind, I felt an inclination to communicate the discoveries I had made during the experiment. I endeavoured to recall the ideas, they were feeble and indistinct; one collection of terms, however, presented itself: and with the most intense belief and prophetic manner, I exclaimed to Dr Kingslake, '*nothing exists but thoughts! – the universe is composed of impressions, ideas, pleasures and pains!*'[41]

Nitrous oxide seemed to Beddoes to offer the prospect of a regulation of mood and spiritual disposition by chemical means, a pledge 'that by ascertaining the action of the elements entering into his composition, Man may, some time, come to rule over the causes of pain or pleasure, with a dominion as absolute as that which at present he exercises over domestic animals and all the other instruments of his convenience'.[42] In 1801 Davy moved to the Royal Institution in London, where he conducted public demonstrations of the gas.

Beddoes's experiments with mental as well as bodily modification by pneumatic means brought scornful mockery from conservative critics. The most extended and damaging example of this was a long mock-epic by Richard Polwhele entitled 'The Pneumatic Revellers', which appeared in the *Anti-Jacobin Review* in 1800. Polwhele dwelt sardonically on the claims made by Beddoes and Erasmus Darwin for the vitalizing and preservative powers of various gases:

> Dr B—— however, combining in his own great and comprehensive mind the theories of Darwin and of Godwin, and applying his dephlogisticated nitrous gas to the purposes of both these philosophers, professes his ability to turn us all into amphibious creatures (as some think, a little out of his own element) – to repair the breaches in our constitutions, whether we have suffered from time or intemperance – to subdue disease and pain – to renovate in the aged every source of pleasure, and even on earth to render man immortal.[43]

His emphasis is on the vanity of theoretical attempts at the unenvironing of man, removing him from the limits of his place and condition. For Polwhele, the air is both the sign and proof of our embeddedness; yet 'aerology', the conquest and modification of the air, are the signs of the human aspiration to disembed ourselves, to take to the air (and indeed to the other elements, taking our air with us). Polwhele made a good deal of the very unethereal nature of Beddoes's own short, stout figure, quoting Beddoes's own account of his experiences of nitrous oxide: 'He seemed to himself, at the time, to be bathed, all over, with a bucket full of good humour.

A constant fine glow, which affected the stomach, led him, one day, to take an inconvenient portion of food, and to try the AIR, afterwards. It very soon removed the sense of distention' ('Pneumatic Revellers', vii).

The poem begins with Beddoes calling his friends to partake of the wondrous gas, which promises to purge and refine all the deficiencies of mortality:

My friends! from a world, where disorders are rife,
I call you, to taste of the liquor of life;
A fluid, to render us nimble and fresh,
And purge from its drossy pollution the flesh.
('Pneumatic Revellers', 1)

Beddoes is then made to proclaim the long ancestry of the gas, going as far as to suggest that gas may have been implicated in the temptation of Eve, though it is clear that its bad effects are the effects of impurity in its composition:

Inciting fond Eve to a spiritual revel,
The very first chemist in air, was the devil.
Yet the substance (alas! we have cause to be serious!)
In Eve effervescing, was damn'd deleterious.
('Pneumatic Revellers', 3–4)

Then, taking advantage of the fact that so many poets were among the subjects whose experiences were reported by Beddoes, Polwhele gives a series of poetic parodies, of the work of George, Mrs Barbauld and Robert Southey, the last providing a rapturous climax:

I spurn, I spurn
This cumbrous clod of earth; and, borne on wings
Of lady-birds, 'all spirit,' I ascend
Into the immeasurable space, and cleave
The clear ethereal azure; and from star
To star still gliding, to the heaven of heavens

Robert Seymour, *The March of Intellect: Poetry*, etching, with watercolour (London, 1829). The caption explains that 'This is not the Laughing, but the Hippocrene or Poetic Gas, Sir, the Gentleman you see inspired here is throwing out the rough materials for an Heroic Poem.'

Aspire, and plunging thro' the sapphire blaze,
Ingulph the dephlogisticated floods
Of life, and riot in immortal gas!
('Pneumatic Revellers', 11–12)

A final song makes the link between Beddoes's vaporous speculations and French revolutionary politics clear.

Then hail, happy says! When the high and the low,
All nourish'd alike from this air-hospitality,
Shall together with gas-born benevolence glow,
And prove, that true bliss must arise from equality;

When, Britons and Gauls! ye shall revel and sing,
(Light, lighter than gossamers twinkle and glance)
Here, thridding a maze, and there link'd in a ring,
And scarcely touch earth, as ye kindle the dance;

When, finer and finer as waxes your nature,
Each atom terrene shall fly off from your bodies,
Each particle gross, and all purified matter,
Ye shall smell of ambrosia or gas like a goddess;

Till mounting, as if in balloons, to the sky,
While pleasure with novel sensations shall strikey,
Thro' the regions of gas shall ye flutter and fly,
A Mercury each man, and each woman a Psyche!
('Pneumatic Revellers', 16–17)

It would be tempting to read the coincidence of the chemical revolution of Priestley and Lavoisier with the political revolutions in America and France in terms of the traditional associations of air, that is, as wild, explosive, ungraspable, capricious. This is how Burke's remarks have usually been read.[44] But in fact the pneumatic revolution marks a new round in the process whereby the air has been brought down to earth, made tractable to practical understanding. The passing of the era of phlogiston was achieved not through a sudden, religious illumination provided by the isolation of oxygen, that elusive, but still long-suspected vital principle hidden in the air, but through the patient and meticulous work of measurement and calculation. The pneumatic chemistry of Black, Priestley and Lavoisier represents a decisive second era – the first having been the work of Galileo, Torricelli, Boyle, Hooke, Mayow and others in the seventeenth century – in the process whereby the air was materialized, drawn in to the world of weight, measure and mechanicity.

Thomas Beddoes had suggested that 'animal motion, at least that of animals analogous to man, would be produced by a very beautiful pneumatic machinery; and our nervous and muscular systems may be considered as a sort of steam-engine'.[45] In a certain sense, this conception dissipates the traditional poetry of air. But we can make out in the delusions of Matthews a new mutation, a kind of pneumo-mechanical poetics, in which the very mechanism of the air provides the warrant for a new kind of enigma and phantasm. Conceived as 'mere' machinery, the mind is reduced to a series of material, mechanical operations. But the particular kind of machine that the mind takes itself for, takes for itself, is also intensely imaginary. The mind is both literalized and refigured by the idea of the machine.

Mesmerism

Matthews's pneumaticians know more than chemistry. They are, Haslam reports, 'superlatively skillful in everything which relates to pneumatic chemistry, physiology, nervous influence, sympathy, human mind, and the higher metaphysic' (*Illustrations*, 57). Roy Porter sees Matthews's delusion as 'a grotesque microcosm of the age', his focus here being not on pneumatic chemistry but on another science, or pseudo-science, which was just as French and therefore also 'a prime target in the politics of panic', namely mesmerism.[46]

Much has been written about this first wave of theories of animal magnetism, as well as about its revival from the middle of the nineteenth century onwards. The tendency in recent years has been to read the mesmeric scene as a displacement of social, and especially sexual relations. The interest of mesmerism has come to consist in its dramatization of operations of power, between operator and subject, doctor and patient, man and woman. Matthews's political paranoia seems like a perfect enactment of such relations of power. To read mesmerism in such terms, as politics rather than physics, is to repeat a move made during the nineteenth century. Mesmer's theory depended on the existence of a subtle fluid that was the medium for the operator's action upon his patients. In 1784 Louis XVI appointed a commission of experts to investigate Mesmer's claims for this fluid. The commission, which included Benjamin Franklin and Antoine Lavoisier, acknowledged that Mesmer's methods seemed capable of effecting cures, but concluded that there was no evidence for the existence of the subtle fluid on which those cures were based. Whatever mesmerism did, it did by means of imagination, not fluid mechanics. The Marquis de Puységur, Mesmer's most important follower, continued to believe implicitly in the physical existence of the subtle fluid (which he identified with electricity), and thought that one proof of the electrical nature of the human frame was that human bones could be ground and heated to produce 'GLASS of a superb transparency' (glass being a material that produced static electricity in conspicuous abundance). Nevertheless, he placed much less emphasis than Mesmer had on material mediation. It was precisely because the fluid was an absolute mediator, since it was present in varying concentrations in all living creatures, that it did not itself require the mediation of objects and apparatuses. Indeed, Puységur was tempted to identify the fluid with thought itself, or at least to represent the power of thought as specially representative of the fluid's power:

> Man, like everything else which exists, is also saturated in its own fashion by the universal fluid and may be considered perhaps as an animal el*ectrical machine*, the most perfect in existence, since its thought, that regulates all his actions, can conduct him towards the infinite.[47]

Most interpreters of mesmerism have amplified or repeated this gesture of dematerialization. But this is to leave out the most important part of the action of imagination in the case of mesmerism, namely that it is a material imagination, which is relayed through properties, apparatuses and substances. Imagination requires the mediation of a material, even if that material is itself imaginary.

The beginnings of Mesmer's conception are to be found in his revival of Richard Mead's Newtonian arguments for the influence of the stars and planets on living creatures in his dissertation *De planetarium influxu in corpus humanum*, presented to the University of Vienna in 1766. For Mesmer, the fact that the movements of the moon produced regular variations, not just in bodies of water but also in air pressure, was a strong indication of the susceptibility of human bodies to atmospheric variations:

> Physico-chemical variations of air, the element in which we live, disturb the harmony of the physical body. Who does not know that the air – hot, cold, dry, humid, in motion, stagnant, rendered foul by various particles – affects all living beings? If the moon can make us be engulfed by an atmosphere raised ten times higher, if it can bring together, from diverse regions, vapors which are scattered over all the horizon and heap them on our necks, if it can then be the cause of winds, heat, cold, clouds, fog, storms, who, I ask, does not see clearly that this star dominates us?[48]

Mesmer postulated an even more fundamental force 'which actually strains, relaxes and agitates the cohesion, elasticity, irritability, magnetism and electricity in the smallest fluid and solid particles of our machine' (*Mesmerism*, 14). This force, which he identifies as 'animal gravity' is not itself atmospheric, since it is itself ultimately the cause of atmospheric phenomena, but seems like an intensification of the idea of an all-pervasive atmosphere. He reported in a *Lettre sur la cure magnétique à un médecin résident à l'étranger* of 1775 that he used magnets to create an artificial tide in the body of a young girl who was suffering from convulsions, vomiting, urine retention, toothache and numerous other maladies. The treatment worked, he thought, not because the nerves were susceptible to magnetism, but rather because 'magnetic matter, by virtue of its extreme subtlety and its similarity to nervous fluid, disturbs the movement of the fluid in such a way that it causes all to return to the natural order, which I call the harmony of the nerves' (*Mesmerism*, 29).

He went a little further in characterising this nervous fluid in a *Discourse on Magnetism* of 1782. Its outward form was to be identified with the many kinds of subtle matter spoken of in occult and scientific tradition:

> Since time immemorial one has spoken of sympathy, antipathy, of attraction, repulsion, of ethereal matter, of *phlogiston*, of subtle matter, of animal spirits, of electrical matter, and of magnetic matter. All these agents, whose action is as real as the existence of light – do they not proclaim the widespread universal fluid? (*Mesmerism*, 34)

The form it took inside the body was the putative 'universal fluid which penetrates us', which was responsible for transmitting our will to our bodies (*Mesmerism*, 34). The therapeutic principle on which Mesmer operates is that this fluid must be maintained in a healthy state of flow. Illness was caused by 'viscosities' in this fluid, which are to be thought of on the analogy of the 'coarse, pasty, viscous moods, produced by bad digestion, [which are] occasioned by congestions and obstructions' (*Mesmerism*, 37). The fantasy here is of a body that is healthy when it approaches the condition of the ethereal, and unhealthy when it is subject to distension, inflammation, engorgement, contusion, slowing, stoppage or solidification. (The ancientness of this way of thinking scarcely needs pointing out.) To be treated by the powers of animal magnetism was to believe oneself to have been converted into the substance that was supposed to be operating on one. It is as though one became well by becoming pure flow, by being subtilized into one's own thought, conceived as an absolutely ideal, because maximally attenuated, matter. This subtle matter is indistinguishably the thought of flow and the flow of thought.

Mesmer characterized the many differing qualities of this subtle fluid in a series of propositions appended to the *Mémoire sur la découverte du magnétisme animal* of 1779. Its most important feature is that it is omnipresent and incapable of being restricted: it is 'a substance whose rarefied nature enables it to penetrate all bodies without appreciable loss of activity . . . Its action is exerted at a distance, without the aid of any intermediate body' (*Mesmerism*, 68). And yet, remarkably, even though it is 'universally distributed and continuous' (*Mesmerism*, 67), this substance can be condensed and stored: 'It is intensified and reflected by mirrors, just like light . . . It is communicated, propagated and intensified by sound', thinks Mesmer. It can also be 'stored up, concentrated and transported' (*Mesmerism*, 68). Thus it is both universally diffused, and differentially concentrated. What is concentrated is the power of dissipation. It is a force which has the powers, and some of the susceptibilities, of a body.

In his later writings, Mesmer enlarged his claims for this fluid's powers to mediate across any and all differences. In *Mémoire de F. A. Mesmer, docteur en médecine, sur ses découvertes* (1799), he made it clear that, as the universal mediator, it has no differentiating qualities of its own:

> Having no particular property, it is neither springy nor ponderous, but is the means in itself of determining properties in all divisions of matter which exist in a more composite form than it does. With regard to the properties which it determines in organic bodies, this fluid is as air is to sound and harmony, or as ether is to light. (*Mesmerism*, 99).

As the medium of universal permeation and communicability, which can overcome every resistance, it is in fact indifference itself:

> . . . there can never occur any movement or displacement, even within its slightest parts, which does not reach, to some extent, the entire expanse of the universe. We can therefore conclude that there is neither a being nor a combination of matter which – by the relations in which they exist in the whole – does not imprint an effect upon all surrounding matter and upon the medium within which we are immersed. (*Mesmerism*, 119–20)

Thus, '[t]he universe is dissolved and reduced to a single common entity' (*Mesmerism*, 99). Finally, Mesmer takes an even more audacious step, one that would encourage the association with spiritualism in the later nineteenth century, by claiming that the universal communication effected by this fluid collapsed time as well as space:

> all which 'has been' has left some sort of trace; similarly, that which 'will be' is already determined by the totality of causes which must bring it about; this has led to the idea that within the universe everything is present, and that the past and the future are nothing but different references which parts of the universe have towards each other. (*Mesmerism*, 122)

Mesmer is at pains to insist that the fluid that transmits animal magnetism is not the air, since '[t]his type of sensation can only be acquired through the mediation of fluids which are as superior in their subtlety to ether as ether is to ordinary air' (*Mesmerism*, 122). But, however super-attenuated it may be, this fluid is still regarded as material, and never crosses the line into pure spirit. As such, its closest material analogue will always nevertheless be the air, which is a matter from which many of the evidences of what is usually thought of as 'matter' are missing, and so the matter that most nearly approximates to the immaterial.

One might be reminded by all this of the fantasy of a 'universal acid', which Daniel Dennett tells us so amused him as a child. The prospect of an acid that is so corrosive that it can eat its way through anything is thrilling and enthralling, especially perhaps to a young boy. But mature reflection throws up a problem: *what do you keep it in*? And, without such a container, how do you keep it from eating its way through the earth and diffusing into space, vaporizing asteroids and other celestial bodies as it goes?[49] Mesmer's universal fluid is a much softer kind of solvent, but it is subject to more or less the same objection. How could it be possible to gather, or store a substance (if it is a substance; perhaps it's a force), whose nature is to permeate all space and everything in it? How could such a fluid possibly be subject to the damming or clotting that Mesmer thought caused illness, for what could possibly constrain or impede it? What could it really mean for a human body to act as a reservoir or conductor of such vital fluid, making it move faster or more freely than normal? How could there be more of it in one particular location than in another (wouldn't this make it denser, less dilute, and thus less able to penetrate bodies and objects?) In fact, like the universal acid, the vital fluid could never possibly be or remain in any one place at all – by the very condition of its nature, it would always be in motion, moving through and between things.

The paradoxical nature of Mesmer's subtle fluid is enacted in the two sides of Matthews's fantasy. There is, first of all, the gloriously imperious dream of universal communication, in which every distance or resistance is set at naught. Although he is the persecuted subject, Matthews is also in some sense the author of this fantasy of what seems like irresistibly diffused, perfectly appeased desire. But, on the other side, there is the belief in the power of the assailers to concentrate, aim and apply differentially this power of universal permeation, which is what leaves Matthews vulnerable to their vicious ministrations. And there is also the fact to be considered of the machine's limited range: Matthews reports that '[s]uppose the assailed person at the greater distance of several hundred feet, the warp must be so much longer directly towards him, but the farther he goes from the pneumatic machine, the weaker becomes its hold of him, till I should think at one thousand feet he would be out of danger' (*Illustrations*, 51). The magical power is constrained by the very source of its power, its subjection to the laws of mechanics.

The Works

This is to say that the air-loom must work, against obstruction, distance and resistance. For machines are there to do work, and the influencing machine is caught up in an economy of work. If the machine is set to work on its victim, its victim must also learn how to labour against it. For Matthews, there are several kinds of work, all of them steadying as well as exhausting. There is, first of all, a labour of vigilance. Matthews must not only stay on his guard against the assaults of his persecutors, he must also go on the offensive against them. His description of the operation of the air-loom's levers is involved and obscure, but seems to involve some kind of battle between him and the operators for control of his breath, and thus of his thoughts.

> The levers are placed at those points of elevation, *viz.* the one nearly down, at which I begin to let go my breath, taking care to make it a regular, not in any way a hurried breathing. The other, the highest, is where it begins to strain the warp, and by which time it becomes necessary to have taken full breath, to hold till the lever was so far down again. (*Illustrations*, 43)

Here Matthews controls his breath in order to resist the action of what he calls 'vital-straining'. By taking control of his breath, he both overrides the automatic process of his normal breathing, and converts himself into a kind of conscious machine, pitting its rhythms against those of the operator and hoping thus to outwit him. There seems to be quite a repertoire of different kinds of lever-work to accompany different tortures, and different responses on Matthews's part to them. His account of the process of what he calls '*lobster-cracking*' seems to involve a complex dance between three dynamic elements: Matthews, the distant operator and the fluid itself, which has its own elasticity:

> [I]n that dreadful operation by them termed lobster-cracking, I always found it necessary to open my mouth somewhat sooner than I began to take in a breath: I found great relief by so doing, and always imagined, that as soon as the lever was at the lowest, (by which time I had nearly let go my breath) the elasticity of the fluid about me made it recoil from the forcible suction of the loom, much in the manner as a wave recoils or shrinks back after it has been forward on the sands in the ebbing or flowing of the tide: and then remains solely upon its own gravity, till the general flux or stress again forces it forward in the form of a wave. Such appears to me the action of the fluid, which, from the time the lever being fully down, loses all suction-force upon it. I always thought that by so opening my mouth, which many strangers, and those familiar or about me, called sometimes singularity, at others affectation and pretext, and at others asthmatic, &c. instantly let in such momentarily emancipated fluid about me, and enabled me sooner, easier, and with more certainty, to fill my lungs without straining them, and this at every breathing. (*Illustrations*, 43–4)

But there is another labour in which Matthews is engaged, beyond the labour of bearing up against the 'assailment' of his tormenters. This is the labour of observation and, allied with observation, of explication and illustration. Matthews is impelled by a duty of unremitting investigation, keeping his tormenters under continuous surveillance, attempting, through his careful illustrations and explications, to expose to public view the nefarious operations of the gang. One might almost say that Matthews is held together by this stern duty of understanding and testimony.

But the machine also demands labour from its operators, who appear to have to give it constant attention. Matthews's representation of the writhing figures of Charlotte and Sir Archy suggests that they may also be subject to some of the painful effects of its warp. What is more, the operators have to struggle even against each other. In his explanation of the process he calls '*Thought-making*', we learn that

> While one of these villains is sucking at the brain of the person assailed, to extract his existing sentiments, another of the gang, in order to lead astray the sucker (for deception is practised among themselves as part of their system; and there exists no honor, as amongst thieves, in the community of these rascals) will force into his mind a train of ideas very different from the real subject of his thoughts, and which is seized upon as the desired information by the person sucking; whilst he of the gang who has forced the thought on the person assailed, laughs in his sleeve at the imposition he has practised. (*Illustrations*, 34–5)

Air/Machine

This paradox is embodied by the idea – the always oxymoronic idea – of the air-machine itself. To understand this, we need to distinguish between two conceptions of space. In the more usual conception, bodies are distributed in space, which is both interrupted and contoured by their presence, as a plain or a wilderness is given form and visibility by the trees, hills and lakes that arise in it. Space is articulated, explicated and orientated by the bodies it contains. Existence in space means that these bodies are determinate: they have a particular size, shape and position, and particular relations with other bodies (they are in contact, at a greater or lesser distance, above, below or beside these other bodies). Bodies mark out places in space which is thereby clumped or quantized – 'striated', in Deleuze's and Guattari's term. In Mesmer's conception, bodies are not distributed in space: rather space, in the form of the infinitely subtle vital fluid, permeates bodies. Distance, location and orientation are inexistent or meaningless from the perspective of this fluid, which nothing can exclude, obstruct or divide, and which allows every part of space and time to be in contact with every other part. The space of this fluid may be thought of as 'smooth' in Deleuze's and Guattari's terms.[50]

The machine is the paradoxical striation of this smoothness, the impossible container for the universal acid. The fluid that the air-loom brews up and spews out, in what appear to be, not merely just blasts or currents, but actual threads or filaments, produces a geometry of spaces. But it is precisely the representability of the fluid and the machine, the paradoxical ability of this fluid to be concentrated, channelled, stopped up, bottled, amplified, which also gives Matthews his power against his persecutors, or the possibility of resisting their power over him. Not just the machine, but the act of illustrating it so carefully may have helped, as Hartmut Kraft suggests, to give a concrete form to Matthews's otherwise formless and deforming agonies of mind and body.[51]

A machine exhausts itself in its operation: its being is its doing. Only a living thing can be more than it does, for instance by meaning to do what it does, or refraining from doing it. To be sure, a machine has potential, indeed it may be thought of as no more than the storing up of such determinate potential. But it has the potential only to perform again what it has already performed in the past. It does not, so to speak, have the potential for possibility that the things we think of as 'living' do.

If, because of its affinity with spirit, air approximates to something like pure possibility, the smooth, unorientated possibility of everything being possible, the air-machine reduces that possibility to particular potentials and protocols – the effects that may be produced at particular times and places and in particular ways. One might say that the mesmerist is already in this sense an air-machine. His techniques and routines, and the apparatuses like the *baquet* that allow them to be exercised automatically, are the ways in which the vital flow can be regulated, by the introduction of differential stresses and tensions into the otherwise absolutely smooth continuum of all-pervasive, resistless fluid. The machine, the machinery, the operator are all kinds of battery – which, in the form of the Leyden jar, had been invented only a few decades before mesemerism.

The air-machine turns quality into quantity, force into substance. It makes the ineffable fluid finite and manipulable. This is because, in its essence, the machine has no secrets, no hidden interiority, no non-functional residuum, no quiddity that is not accounted for in the details of its operation. To be sure, there are complex, inefficient and exhibitionist machines, machines which seem to consist of show; but in such devices, these are inessential elements, which do not belong to the machine itself. The essence of the machine is that it is has no essence separate from its action, and is thus finitizable, totalizable.

But there is also a kind of wild or infinite machine, which exists as a generalized and proliferating machinery, a machine without a limit, that propagates rather than exhausting itself in its operation. This kind of machinery does not coincide with what it does: it includes and exceeds what it does. Machines are demonstrative in their nature: they can be opened out and revealed. But the air-loom is a black box: it cannot be seen.

The infinite machine is at once a desublimation and a remystification of thought. Conceived of as 'mere' machinery, the mind is reduced to material, mechanical operations. But the growing power invested in the idea of the machine, which increasingly could be thought of, not just as supplementing human actions, but as displacing them, gave it a new, quasi-magical autonomy. The more that thought became automatic, the more autonomous that automaticity could become. The air is increasingly subject to mechanics – but, in Matthews's imagination, the machine had itself already begun melting into air.

Inheritance

Comments have been offered on the importance of the air in Matthews's conception of the air-loom, and, to a smaller degree, of the loom. But Haslam, Porter and Jay all pass over in silence the modification of 'heirloom' that gives his delusive machine its name. What might this mean, beyond furnishing a handy homonym? What inheritance might be represented by the machine? What kind of inhering are we to hear in the air-loom?

We know almost nothing about Matthews's personal past, or his family life, unlike other celebrated paranoiacs. Perhaps there is some private play with the suspicion that his condition is hereditary, and that he has thus inherited the fiendish and yet slightly wonky mental apparatus of the air-loom instead of a smoothly functioning brain. But we know nothing of the familial ills that Matthews may have been heir to. If we even knew something about Matthews's father and Matthews's relationship to him, this might provide the basis for a family romance, which would perhaps allow us to speculate about what Matthews may have inherited from his father, or in what sense he was his heir.

But Matthews, like his machine, seems to have started up out of nowhere, to have no identifiable forebears or antecedents. If he had descendants and if they know who they are, they have yet to come forward. We do not learn from the material that Haslam furnishes in his *Illustrations of Madness*, when the air-loom (one of many) was constructed or how long it had been in existence. We have got into the habit of using the evolution of technology as an index of the passing of time – with the iron age, the steam age, the space age and so on. Matthews's machine, by contrast, seems strangely unlocated. It is hugely sophisticated, employing the newest kind of technology. Matthews, as reported by Haslam, assures us that 'whenever their persons shall be discovered, and their machine exhibited, the wisest professors will be astonished at their progress, and feel ashamed at their own ignorance. The gang proudly boast of their contempt for the immature science of the present era' (*Illustrations*, 57). But the machine is also archaic, with its barrels of noxious substance from which the assailing airs are brewed, and the strange windmill contrivances that are supposed to waft the magnetized effluvium in the right direction.

It is similarly unclear when Matthews began to incubate his delusion. In the highly coloured dramatization of Matthews's visit to the House of Commons in December 1796 which opens *The Air Loom Gang*, Mike Jay represents Matthews as already struggling in the magnetic toils of the machine.[52] But, in a portion of Haslam's memoir, given in Matthews's own words, we discover that 'The assassins opened themselves by their voices to me about Michaelmas 1798' (*Illustrations*, 59). In his own interpretation of the machine Jay follows Roy Porter in seeing it as a dramatization above all of his own condition in Bedlam (*Illustrations*, xxxvii–xl), which might also suggest that the delusion matured and solidified only after 1796, and perhaps quite some time into his confinement. But we cannot be sure either that Matthews's delusion remained in full force in 1810 – it might well have suited Haslam's purposes to represent it as in full flood, even though Matthews seems to have given no sign of his consuming obsession with the air-loom in his meetings with Drs Birkbeck and Clutterbuck during the latter part of 1809.

An heirloom is a possession, but one that is in permanent passage, passed on across generations, but always within a family. Matthews passed his air-loom on to Haslam, perhaps hoping for wider dissemination, and indeed Haslam did not keep it in the family, but passed it on to his legatees, who include the lineage of other victims of different kinds of influencing machine: Friedrich Krauss, John Perceval, Daniel Paul Schreber, Victor Tausk. Indeed, Matthews seems to have set in motion a subtle, insidious contraption that would come to seem like a perpetual motion machine, without beginning or end, the ur-machine of persecutory delusion that, like Joseph Conrad's nightmarish 'knitting-machine' of blind autocreation, 'evolved itself . . . and behold! – it knits . . . It knits us in and it knits us out', and, because it perpetually gives rise to itself, seems to belong securely to no time.[53]

Ingenious engines of thought appear in the delusional systems of other nineteenth-century madmen. None of them seems as fully developed or autonomous as Matthews's, nor do they seem to have been visualized as clearly and continuously as the air-loom. They function as machinery in a more general sense, rather than as single machines. But perhaps they can nevertheless be seen as adjuncts or outworks to Matthews's machinery. And all these machineries have a pneumatic component, though it may be less obvious or central than in Matthews's.

Life-Ether

The first of Matthews's heirs of whom we know was Friedrich Krauss. Krauss was born in 1791 in Göppingen. After a distinguished academic career, he was appointed to the firm of Daniel Thuret and sent to Antwerp in 1814, where his delusions of persecution began in 1816. After a period of confinement in the Cellitenkloster Institution in Antwerp, he was returned home to Göppingen. Here, he wrote the first of his many petitions to the German and Belgian authorities, begging them to take measures against his persecutors. From 1819 to 1824 he gave classes in Heidelberg in languages, commercial law and technical chemistry. From 1827 he resumed his occupation as a travelling commercial representative. In 1832 he began keeping a diary in which he detailed his various mental torments. These were to last his entire life, and to form the subject of two long autobiographical accounts, *Nothschrei eines Magnetisch-Vergifteten* (*Cry of Distress By One Magnetically Poisoned*) (1852) and *Nothgedrungene Fortsetzung meines Nothschrei* (*Enforced Continuation of My Cry of Distress*)(1867), written only some years before his death. There is only one surviving copy of the first volume, in the library of the University of Tübingen. The library of the Society for Psychical Research holds the only surviving copy of the latter. Only selections from the texts have been made available in modern editions.[54]

Krauss wrote of his torment by a number of voices, which soon coagulated into characters. During the early phase of his persecution, he tried to kill himself by dashing his head against the wall; but, as in other such cases, the more the agencies became persons, the easier it seems to have become to cope with them, as his anxiety turned into outrage (*Notschrei*, 19).

There were four principal persecutors, whom he called The Old Magnetizer, Janeke Simon-Thomas, Van Asten and the daughter of Van Asten. They could read his thoughts (leading him to call them 'soul-worms' and 'thought-vultures', *Nothschrei*, 14). They also exercised control over his feelings and moods, producing depression and numbness, as well as distorting his senses of hearing and touch and inflicting actual physical torments. At times, he felt distinctly that his vocal apparatus had been completely taken over and directed according to the will of his persecutors (*Grenzgänge*, 163). During all this time, Krauss struggled to maintain his professional and social position, even though his tormentors would constrain him to perform violent and inappropriate actions, such as leaps into the air, while he was trying to conduct business.

Since he remained uninstitutionalized, Krauss was free to seek information regarding the theories of magnetism and mesmerism as well as professional advice from scientists and physicians, some of whom confirmed his belief that he was subject to magnetic influence, but were unable to assist him. He believed that the publication of his work would encourage others subject to the same torments to come forward, and suggested that doctors had told him of other cases known to them of persecution by animal magnetism (*Grenzgänge*, 41). He seems to have known of Matthews's account of his illness, if only through hearsay or repute (*Nothschrei*, 25). Thirty pages of the 1852 *Nothschrei* were given over to correspondence between Krauss and a Herr L'Hermet of Magdeburg, who also believed himself to be magnetized. L'Hermet had managed to keep his auditory hallucinations under control, so that he heard at most 'a light hissing or sizzling' (*Nothschrei*, 23), though bodily hallucinations continued strongly, for example, of 'numbing ether-stuff pouring into the ears' (*Nothschrei*, 23).

Krauss believed that the material means whereby his persecution was affected was magnetism, which, like many others, he imagined as a form of electricity. Later he would come to refer to it as 'Life-Ether', '*Lebensäther*' (*Nothschrei*, 14). Although animal magnetism resembles electricity, it has an advantage over it:

> As is well known, electricity is transmitted only as long as it is in contact with the wire or with the person who is holding it. By contrast, the electromagnetic fluid maintains its course like a fishing line through the body affected by the magnetiser; once that body has been attracted by the discharge of the apparatus, a flow of influence is guaranteed through all circumstances, and for as long as the magnetiser wishes. (*Grenzgänge*, 50).

Krauss too believed himself to be subject to ear-entry by the malignant gases. He explains that '[t]he ears are the channels to which Nature has given the office of . . . drawing out the magnetic fluid from the air of the world, taking it in and introducing it into the body . . . It is for this reason that this concentrated ether entered me through the ears' (*Grenzgänge*, 181). Krauss believed that the magnetizers used various carriers or media to transmit the fluid, but in particular gases. At one point he distinguishes carefully between three different forms of magnetic gas:

> From the beginning, I distinguished three . . . distinct types of magnetic gas: 1) the usual, that flowed in with only a weak sound, like the hissing of boiling water 2) that which steamed in with a loud sizzling and grating, like the scraping of sand, and powerfully touched, stretched and suffused nerves, muscles and veins . . . 3) the most concentrated, densest and most powerful gas . . . This was drawn out in a very high-pitched sound, like hoooo or tsiiiii . . . like a living flame that reached in terribly, stimulated everything to its highest pitch, kindled everything in an instant and caused me the greatest pain, the most fearsome agony. (*Grenzgänge*, 120)

Krauss suggested that the magnetizers used machines in order to transmit their streams of magnetic influence to him. Though he has clearly read widely in the theories of animal magnetism, citing in particular D. G. Kieser's *System des Tellurismus* of 1826 (*Grenzgänge*, 46–8) and has plenty to say about them, he is much less interested than Matthews in determining or explicating the nature of these machines.[55]

It is difficult to be sure whether Krauss's machine is less developed than Matthews's or more diffused. Krauss emphasizes that the secrets of electromagnetism are spreading ever wider, and taking ever more various forms. The widespread use of electricity meant that 'what was once thought to be a secret concealed by a few specialists and so-called black magicians, now lies so open to view, that the multifarious applications of this power are becoming ever more general' (*Grenzgänge*, 50). This meant that

> [t]his animal or electromagnetism, also known as tellurism, as the mightiest ethereal power, as Life-Ether, has immeasurable reach;

> and, through the continued researches of brilliant intellects, has expanded into the sphere of the incredible, and may easily bring about a transformation in spiritual matters greater than that effected by steam in material life. (*Grenzgänge*, 37)

Animum Rege

John Perceval was the son of Spencer Perceval, the prime minister who was assassinated in the lobby of the House of Commons in May 1812, when his son was aged nine. John Perceval joined a cavalry regiment and, after seeing service in Portugal, rose to the rank of captain. In 1830 he left the army and travelled to Scotland, where there was an outbreak of charismatic Christianity, characterized by speaking in tongues and other ecstatic religious appearances. Already, Perceval was beginning to behave in strange ways, finding himself impelled – or permitted – to break into religious utterance. He travelled from Scotland to Dublin, where his behaviour became so very disordered that he had to be restrained in a room of the inn where he was staying. In December 1830 his brother came to fetch him, and took him to an asylum run by Dr Fox near Bristol. He was confined in this institution for eighteen months and then transferred to a second institution, at Ticehurst in Sussex, run by C. Newington, where he remained for almost another two years. After his release, he wrote and published anonymously *A Narrative of the Treatment Experienced by a Gentleman, During a State of Mental Derangement* (1838). A second edition of the work, which adds much material and amplifies and particularizes the complaints of his treatment in Dr Fox's institution, appeared with his authorship openly acknowledged in 1840. The two books were edited into a single volume by Gregory Bateson in 1961.

Thereafter, Perceval began an active and life-long campaign against the injustices of the system of incarceration and care of lunatics. Perceval has become something of a hero of the struggle against the cruel and brutal treatment of the mentally ill, though many of his complaints about his own treatment revolve around his indignation that a gentleman such as he felt himself to be should have had to endure the indignity of being banged up with his social inferiors.

Perceval's delusions had a strongly religious cast, and consisted very largely of the hearing of voices, which issued a bewildering variety of demands, exhortations and imprecations. But the theme of machinery also runs through his account, though, as with Friedrich Krauss, the machinery of his delusions is perhaps less apparent than in Matthews's because of their predominantly auditory nature.

Perceval seems to have experienced an extreme alienation from his own body, which he represents as acting and acted upon as though it were mechanical. At various times, he feels his body moved automatically, as though by some mechanism, as, for example, when he strikes his neighbour at the tea table: 'My hand struck that blow, but it was involuntary on my part, as if my hand had been moved by a violent wind. A spirit seized my arm with great rapidity, and I struck as if I was a girl.'[56] It is unclear whether this machinery is pneumatically or electrically driven. On another occasion, his right arm 'was suddenly raised, and my hand drawn rapidly across my throat, as if by galvanism' (*Perceval*, 118). The most explicit indication of his mechanical sense of his body is his claim that 'My loss of all control over my will, and belief, and imagination, and even of certain muscles, was immediately preceded by three successive crepitations, like that of electrical sparks in the right temple, not on the same spot, but in a line, one after the other, from left to right' (*Perceval*, 284n). Sometimes Perceval combines the mechanical with the organic, as when he characterizes two painful sensations from which he seeks relief in loud singing: 'the one, at the back of the palate, consisted of a dull, heavy impression, as if made by a thick, mucilaginous spittle – the other was more painful, and about the top of the throat, as if the breath came up very fiery, and impregnated with electrical matter' (*Perceval*, 286). Perceval seems to have associated the actions of eating with the hearing of voices, and his voices seem to have been stimulated particularly at mealtimes. Indeed, he explains the visions and admonitory voices experienced by mystics by reference to the stomach: 'The mind was prepared for receiving the commands supposed to be divine, by the castigation of the stomach, with which the nerves of the brain are so intimately connected' (*Perceval*, 298).

Perceval also seems extremely sensitive to the effects of air and wind, especially as they relate to temperature and to voice. His madness was precipitated, he believes, by an accumulation of causes, including shame, ingratitude and remorse, combined with physical causes of a surprisingly banal nature:

> . . . the abiding presence of this guiding power influencing my actions, and awing my mind, added to the subtle effects of the mercury upon the humours of my body, during the use of which I had the imprudence to expose my frame to draughts, whilst washing for a long time, every morning, my person in cold water, at that inclement season of the year; these causes all combined, could hardly fail to effect the ruin of my mind. (*Perceval*, 26)

Perceval seems to have suffered from an extreme alienation from his own thoughts and process of thinking, an alienation that makes him unable to recognize his thoughts and thought processes as his own. But this alienation also seems to give him occasional glimpses of the process or machinery of his thinking, though he is much less precise than Matthews about the nature of the machinery involved. His alienation thus appears to bring him close to consciousness of his own thought:

> I recollect I found myself one day left alone, and at liberty to leave my bed. I got up, and knelt down to pray. I did not pray, but I saw a vision, intended, as I understood, to convey to me the idea of the mechanism of the human mind! (*Perceval*, 54)

There are hints scattered through his testimony that this machinery is at least in part pneumatic. Perceval sees many of his mental disorders in terms of disturbances of the air and breath:

> I have found that whenever my bodily health has been deranged, particularly whenever my stomach has been affected, I have been more than usually troubled by these fancies, particularly if at the same time, through sluggishness or through cold, I have not been breathing through my nostrils, or drawing deep breaths. (*Perceval*, 298)

One of the strangest of these disturbances is what might be called *panophonia* – the production of voices out of ordinary sounds, especially the internal sounds of his own body: 'I found that the breathing of my nostrils also, particularly when I was agitated, had been and was clothed with words and sentences' (*Perceval*, 295). The sound of air is particularly liable to become, in his expressive phrase, 'clothed with articulation' (*Perceval*, 265). He describes his fear at the approach of his attendants: 'Their footsteps talked to me as they came up stairs, the breathing of their nostrils over me as they unfastened me, whispered threatenings; a machine I used to hear at work pumping, spoke horrors' (*Perceval*, 93). As he begins to recover, he is able increasingly to identify the sources of these sounds:

> I discovered one day, when I thought I was attending to a voice that was speaking to me, that, my mind being suddenly directed to outward objects, – the sound remained but the voice was gone; the sound proceeded from a neighbouring room or from a draft of air through the window or doorway. (*Perceval*, 294)

On another occasion, gas-jets are identified as the source:

> Continually over the head of the bed, at the left-hand side, as if in the ceiling, there was a sound as the voice of many waters, and I was made to imagine that the jets of gas, that came from the fire-place on the left-hand side, were the utterance of my Father's spirit, which was continually within me, attempting to save me, and continually obliged to return to be purified in hell fire, in consequence of the contamination it received from my foul thoughts. I make use of the language I heard. (*Perceval*, 45)

Elsewhere, he reminds his readers that 'the ancient priestesses at Delphi and elsewhere inhaled a powerful gas before giving their oracles' (*Perceval*, 327). He continues to attribute this hallucination to the effect of a power of some kind exercised from without or within, either by God or by some agency within him more aware of the operations of his mind than he.

If voices are produced through disturbed respiration, then, he believes, recovery will entail the proper regulation of the machinery of breath: 'I question whether the operations of the conscience and reflection can be conducted but through the medium of the lungs filling the chest at proper intervals, according to the degree of passion of the mind, or of action of the body' (*Perceval*, 272). He believes that this disordering can be reversed by conscious control of the breath:

> I believe the healthy state of the mind depends very much upon the regulation of the inspiration and expiration; that the direction '*animum rege*', has a physical as a well as a spiritual sense; that is, in controlling the spirit you must control your respirations. I will instance, in support of this, the stupid appearance of many deaf people, who are usually unable to breathe freely through the nostril, and keep their mouths wide open; a habit very common among idiots. I will instance, again, the stupefying effects of a bad cold. (*Perceval*, 271–2)

Perceval suggests that mechanical assistance may help to regulate the breath: 'it is possible, that the effecting of this [breathing] mechanically even may give much relief. I have certainly found it so' (*Perceval*, 272–3).

Elsewhere, Perceval defines madness as the incapacity to distinguish literal from figurative expressions, and the consequent tendency to take metaphors literally. This applies particularly, he thinks, to the voices he heard commanding him to suffocate himself: 'when I was desired to suffocate myself on my pillow, and that all the world were suffocating for me, &c. &c., I conceive, now, that the spirit referred to the suffocation of my feelings – that I was to suffocate my grief, my indignation, or what not, on the pillow of my conscience' (*Perceval*, 271). But his emphasis on the literality of breathing, on the physical as well as spiritual sense of '*animus*', indicates that he is still inclined to take the letter for the spirit, especially when it is the letter of the spirit: for Perceval, thought is, more than ever, breath. This conviction of the literal link between thinking and breathing connects with his curious obsession with suffocation. One of the earliest signs of his madness was a compulsion to try to suffocate himself in his pillow – 'I imagined that I should be really suffocated, but saved from death, or raised from death, by miraculous interposition' (*Perceval*, 43). He suspects his keepers of trying to suffocate him and, alternatively, of inducing sudden and violent panting by plunging him into cold baths and showers, a practice that he suspects indicates that the secret link between mentation and respiration is known to his keepers (*Perceval*, 272).

Perceval sums up his beliefs in a series of propositions:

> To make my ideas more clear, let me sum up my arguments or propositions thus: That a healthy state of the mind is identical with a certain regulated system of respiration, according to the degree of bodily action; that the exercise of reflection or of conscience, in the control of the passions or affections of the mind, is concomitant with, or effected by a proper control of the respiration – quiet when the mind is quiet, accompanied with sobs and sighs when otherwise. That the mind and the blood, being intimately connected, the health of the body depends also on this healthy regulation of respiration, promoting a proper circulation and purification of the blood; that, consequently, the effecting respiration by mechanical means, without the control of the muscles by thought, is profitable to the health of the body, and also to that of the mental faculties, although they may not be, at least distinctly, occupied by any ideas; in the same way as, if several printing-presses are worked by machinery, it may be necessary for the perfect state of that machinery, that all the presses should be in motion, though some may have no types under them. (*Perceval*, 273)

This statement hovers in a characteristic way between conventional wisdom regarding the coordination of breathing and cognition and obsessional literalism. Perceval seems to imagine a kind of iron lung or automatic regulator of the breath to ensure the orderly production of thoughts – here imagined as the output of a printing press. Remarkably, he seems to have no sense of any connection to the import of these thoughts, to which he stands in the same relation as the proprietor of a printing works to the content of what is being printed – and Perceval is drawing attention to the fact that the machine needs to be kept running even when there is no type to be reproduced, or thoughts to occupy the mental faculties. This is the maturing of Perceval's image of his mind as a kind of pulmonary thought-works.

In one passage, he speaks of the power that can be released when judgement is suspended or overturned, a power that is displayed in the inspiration of 'many poets, many painters, many singers, many actors, and even orators [who] have never spoken, acted, sung, designed, or written so well as when they have been intoxicated'. But the 'passion and feeling' which produce 'splendid diction, splendid action, splendid delineation' are not a transcendence of the merely mechanical faculties of the body and mind, so much as an expression of the 'excellent machinery' of the mind:

> I think, therefore, that by the observation of the operations of the mind, under such circumstances, much spiritual and even *physical* knowledge may be obtained, because I am convinced that the mind is a piece of excellent machinery. Like to a musical instrument, whose movements we are yet to discover how to regulate, by certain fixed and, if I may call them so without offence, mechanical laws. (*Perceval*, 283)

Perceval's delusive states, like those of Matthews and Krauss, are characterized by an extreme somatizing, in which cognitive and moral processes are displaced into physical processes and functions, in a kind of mechanization of thought. Perceval experiences his own thought as a complex and unpredictable machinery. His recovery comes about, he assures us, when he realizes that his voices are speaking allegorically or figuratively, his madness consisting in taking metaphors *au pied de la lettre*. But these mechanical processes persist in his explanation of how the illusion of voices is produced, by a process of what he calls the 'clothing with articulation' of ordinary sounds. He also provides an extraordinary explanation of the process whereby visual hallucinations were produced in him, in part by causing the phosphorus that he believes predominates in the brains of lunatics to flash. Perceval offers us as a proof of his recovery his recognition that the pictures he saw were not real, but phantasms produced in him 'as the spectre is thrown out of a magic lantern' (*Perceval*, 306). But he attributes this operation to the Almighty. And why should the Almighty disorder his senses in this way? Just for fun: 'Is God a God of the sincere, the grave, the sober, and the chaste only? Is he not also the God of fun, of humour, of frolic, of merriment, and of joviality?' (*Perceval*, 311). This practical joker of a God seems to be a version of the influencing machine.

Tausk's influencing machine is the body, and more specifically the genitals, made unrecognizable by the machinations of desire, power and fear. Presumably some part of the therapy and recovery of the psychotic would be the unmasking of the body behind the machine. Perceval's recovery also takes the form of a demystification, or a somatizing, of effects that had been thought to be spiritual or supernatural. Voices are explained in terms of the mechanics of auditory hallucination; the exercise of sound judgement is associated with the equipoise of the breath. But Perceval's body is not a deliverance from the obfuscations of the machine, or from the machinery of his delusion, for his is a mechanical body, which is to take the place of the spiritual body of which he believed himself possessed at an early stage of his madness. Perceval sees the understanding of the machinery of his body as the proof of his recovery, when it is in fact the most tenacious aspect of his pathology. The neutral mechanism of his body and breath may take away the authority of his voices and allow him to ignore rather than strive to obey them, but that mechanism is still deeply delusive. In Perceval's case, the turn to the body cannot wholly dissipate the mechanism of his delusions, since his body is a delusive machinery. We should remember that, whereas Matthews's air-loom belongs to the most florid period of his persecutory delusion, these and other theories belong to the period of what Perceval insists is his recovery. Perhaps an externalized influencing machine never materialised in Perceval's thinking because his influencing machine is, in the end, the ideal, regulated mechanism of his own embodied thinking, which he is convinced has delivered him from the disordering of his wits. Recovery, for Perceval, seems to consist in replacing the idea that he is possessed by spiritual agencies with the idea that he is suffering from a disorder of his mental machinery – but the idea he entertains of this mental machinery is subtly, but enduringly delusive. In Matthews's case, the consolidation of the influencing machine was both the fulfilment and the containment of his madness. For Krauss and Perceval, the influencing machine is subdued, not by being destroyed, but by being generalized. Perceval in particular overcomes his influencing machine by becoming it.

Because air is the privileged matter of thought, the machinery of air is a way for thought to think its own workings, encouraged by the development of machines that produced, manipulated and projected across the spaces of the air. The fluid mechanics of air provide a way of embodying the complex machineries of thought, especially in its self-attention. As the fugitive matter of mind, air had always been a way of imaging the irreducibility of mind to body, and even the intractability of mind to itself. But the delusions of mechanoiacs – mechanical paranoiacs – are not so much the signs of a dissolution of a Cartesian subjectivity by telematic media, or of the eruption of 'the unspeakable reaches of his unconscious', as Mike Jay suggests,[57] as the signs of a crisis of hyperconsciousness, a consciousness brought to crisis by the terrifying intensity of its fancied consciousness of itself. So the problem is not one of being alienated from one's thoughts, or dispossessed of them, but rather of coming too close to them. The passion of the machine arises not in the fear of alienation from it, but in the grimly voluptuous delight of identification with it.

Indeed, the apparent terror at possessing agencies may be a systematic defence against the greater terror of having no other way to escape from this excruciating proximity to self. The fantasy of the machine out there is a repudiation of the machine of fantasy in here. It allows its victim to say 'I am assailed', 'I am worked', 'I am possessed', to preserve the 'I' in the agonized cry 'I am not myself'. It is a defence against the omnipotence of thought, the fearsome identity of thought with itself and the dominion of thought over itself (which means the subjection of thought to itself). It is a problem of reflexivity, the pathology of persecutory self-knowledge. It is for this reason that there is so close a relationship between sickness and therapy in such conditions, both of which take the form of laying bare the machinery of the soul in order to allow the self to exercise mastery over the machine. The influencing machine is a desperate attempt to restore the world, to restore the otherness of the world. Matthews is not trying to save his soul, but to save himself from it. For what does it profit a man if he gain his soul but lose the whole world?

So there must be two kinds of air-machine. One is bulky, odorous, corporeal, kinetic. This is set against another, which is volatile, edgeless, self-generating. It is in fact a bad infinity, part of the infinite machine, or dissolution of the world into thought. It is the other side of Davy's enraptured pneumatism (*Nothing exists but thoughts!*), in the dominion of mechanical spirit. This is not a defence of the infinitude of the air against its finitizing, but a defence of the concrete against the abstract air, the air of breath against the air of universal thought. This explains the desire to materialize, or mechanize the self, in archaic machines that are bulwarks against the infinite machinery of thought. Matthews, Krauss, Perceval and Schreber do not want to be integrated with their thoughts they want to be smaller than their thought. The thought 'nothing is external to me' is made over into the thought 'everything is external to me.' The fear of becoming everything is disguised and deflected by the fear of becoming nothing. The archaic machine is a body in space, which explains the importance of the white space in Matthews's drawing. As the influencing machine is elaborated, it will expand to occupy more and more of the available space – pictures by other sufferers from a later period show a much more saturated space, full of wires and interconnections. Matthews's machine is incomplete, and he has directed that some parts of the illustration be left sketchy. This may suggest, both that the machine has not yet been fully revealed to him and that the machine itself is dissolving into the air that it itself weaves – as though the machine were indeed beginning to become everything, to take to the air. The yawning spaces of the illustration, and the incompleteness to which they testify, are a way of keeping himself apart from himself, of forestalling his identification with, and as, 'Bill the King', the leader of the gang, who is nowhere to be seen in the illustration.

The mechanization of air doubles and assists the corporealization of thought. But the resulting body of thought is an air-body, made material in the image of what was beginning to become the reference-state of matter – that is, the volatile, the vaporous, the airy. In the process, the machine, too, changed its character. It became generalized, indefinite, ubiquitous, self-generating. These men came early upon a machinery that had become compounded with air.

three

Inebriate of Air

Thinking About Thinking

Freud identifies magical thinking with what he calls 'omnipotence of thoughts'. He believes that this attitude is a return to the 'animistic' phase, which is characterized, as he puts it in 'The Uncanny' (1919),

> by the idea that the world was peopled with the spirits of human beings; by the subject's narcissistic overvaluation of his own mental processes; by the belief in the omnipotence of thoughts and the technique of magic based on that belief; by the attribution to various outside persons and things of carefully graded magical powers, or '*mana*'; as well as by all the other creations with the help of which man, in the unrestricted narcissism of that stage of development, strove to fend off the manifest prohibitions of reality.[1]

The phrase 'omnipotence of thoughts' is not Freud's own coinage, but that of the patient he called the 'Rat Man'. As he explains in *Totem and Taboo* (1913), the Rat Man believed in his capacity to affect reality through his thoughts:

> He had coined the phrase as an explanation of all the strange and uncanny events by which he, like others afflicted with the same illness, seemed to be pursued. If he thought of someone, he would be sure to meet that very person immediately afterwards, as though by magic. If he suddenly asked after the health of an acquaintance whom he had not seen for a long time, he would

> hear that he had just died, so that it would look as though a telepathic message had arrived from him. If, without any really serious intention, he swore at some stranger, he might be sure that the man would die soon afterwards, so that he would feel responsible for his death.[2]

As the belief that thoughts have the power to affect the world, indeed, as the mistaking of one's thoughts *for* the world, magical thinking might not seem at first sight to cover the case of superstitious, ritual or magical behaviours, in which actions rather than thoughts seem to be to the fore. After all, a magical action – walking under a ladder, breaking a mirror – may well be thought to have consequences that are entirely separate from the thoughts that one might have about them. We may suggest, however, that in reality such actions are a species of thought-in-action, in that they belong to and confirm a magical theory of the world that represents a triumph over it, a subordination of the world to thought, only here made autonomous of thinking itself.

Freud may inherit something of J. G. Frazer's admiration for, and even partial identification with, the magical stage of human development. It is plain that, although Frazer regarded the magical stage of human thought as more primitive than the religious, he respected the rationality of magical thinking, cracked rationality though it might be; for at least magical thinking posited a world of invariant physical laws applying indifferently to all forms of life, physical and spiritual, as opposed to a world governed by the randiness, rancour or toddler-like tantrums of supernatural beings. Magic, like science, depends upon the power of thought both to apply and subordinate itself to law.

Freud suggests that in the animistic period (religion without gods), man had untroubled confidence in himself and his powers. This idea recurs throughout psychoanalytic writing and psychoanalytically inflected ethnographic writing on magical practices and procedures. René Spitz sees the move from the magical to the religious stage as involving the idea of transcendence, with its associated requirement of a split between the subject and the object. By contrast, he affirms,

> magical thinking knows nothing of transcendence. In the beginning, primitive man does not sense any difference at all between subject and object. Later, to be sure, he objectifies the external world, but he himself is on the same footing with it. Even in the most highly developed magical cults, the basic principle remains the possibility of *coercing* the godhead.[3]

But what happens when thought's omnipotence turns on itself? In more recent times, magical thinking has become reflexive, involving not only the power of thought over the world but also the coercive power of thought over itself, along with (the thought of) the power of others' thoughts over one's own. Freud does not neglect this question entirely. Indeed, he derives the feeling of the uncanny from an uncomfortable return of the idea of the omnipotence of thoughts, as though the idea of omnipotence represented some threat to the more modest, constrained attitudes of 'the scientific view of the universe', which 'no longer affords any room for human omnipotence' and in which 'men have acknowledged their smallness and submitted resignedly to death and to the other necessities of nature'.[4] The uncanny is thus the result of a kind of interference of different modes or epochs of thought.

Perhaps the most striking thing about magical thinking is that, for the most part, it does not, as Freud and Frazer might have wished, constitute omnipotence of thoughts. Magical thinking is typically expressed in habits, rituals, obsessions, in which, far from being free to exercise its dominion over the world, thinking is subjected to austere and inexorable regulation. Magical thinking may be powerful, but it is by no means free. Magical thinking seems, in fact, to be based less on free omnipotence of thoughts than on a complex and ambivalent bringing together of power, passivity and persecution. Some psychoanalytic case-histories – for example, Theodor Reik's 'On the Effect of Unconscious Death Wishes' (1913) – have stressed the alienating effect on patients who suspected themselves of being able to control reality by their thoughts, and regarded themselves as the helpless victims of that power.[5]

Magical thinking is usually mechanical, reminding us forcibly of the shared etymological roots of magic and machinery. Nowhere is this more the case than when thinking itself is caught up in the magical economy of power. As we saw in chapter Two, there is reason to suspect that, from the eighteenth century onwards, thinking indeed came more and more to think of itself in terms of machinery – the subject as Rochester's 'reasoning engine'. Accepting Freud's definition of magic as omnipotence of thoughts, but concentrating on the magic of thoughts rather than the thought of magic, I want in this chapter to consider the ways in which thought strives and is constrained to exercise its coercions on itself during the later nineteenth century. Where Chapter Two registered the pressure of gaseous forms on the self-conception of the mad, I will here track some of the appearance and effects of gas and air in more philosophical forms of thinking about thinking.

A Matter of Thought

Despite enduring philosophical rumours to the contrary, it is not very easy for human beings to entertain an immediate apprehension to their own thinking. 'The more I think, the less I am', as Michel Serres has percipiently observed.[6] Thinking of thinking is a magical operation, in that it requires a subject to constitute itself as an object, while not surrendering its subjective privileges. For thinking to make of itself an object, it seems to need a material substrate, which may be defined as that which underlies, or provides a ground or foundation for, an otherwise immaterial thing. There is one particular substrate for the immaterial action of thought which has had prestige across many cultures and languages, namely the air and, from the end of the eighteenth century onwards, its many gaseous correlates. Like thought, the air has power without presence. It is invisible, but has visible effects. Oddly, then, in serving as the substrate for thought, the air succeeds precisely to the degree that it does not in fact function as a substance, but rather a quasi-substance, a substance that, like thought itself, is next to nothing, not quite there. Air is a thought-form. It is a substrate for that which has no evident substrate. Air is always more or less than air: more because it is always in some measure the idea as well as the simple stuff of air, less because it is never fully present in or as itself, and so only ever air apparent. In this non-adequacy to itself, it resembles the thought it figures, and is thus at once adequate and fittingly inadequate for that figuring. For this reason the powers of air are not just imagined or imaginary; they are materializations of the power of imagination itself.

Trances of the Blast

From the end of the eighteenth century, advances in pneumatic chemistry made the old affinities between air and thought both more various and more accostingly actual. The late eighteenth century saw a sudden breakthrough in the understanding of the chemistry of gases, with the discoveries of Black, Cavendish, Lavoisier and Priestley, and the isolation of oxygen, carbon dioxide and other gases. Following the work of Beddoes and Davy, oxygen, the principle of life, was invested with huge therapeutic powers, with air-baths and other therapies being vigorously marketed and the whole century was becoming, as Emily Dickinson declared herself to be, 'inebriate of air'.[7] This belief in the vitalizing powers of air broadened into the Victorian cult of ventilation and fresh air. The nineteenth century saw a struggle between an older fear of exposure to the air and campaigns to promote more vigorous and healthy forms of that exposure. On both sides of the dichotomy we may detect a magical aggrandisement of the powers of air, as the bringer either of health or of infirmity. Air-thinking exhibited a characteristic compounding of omnipotence and anxiety, which is nowhere more marked than in the superstitious Victorian fear of draughts and specifically the belief that exposure to small currents of air will cause illness, a mild and magical delusion that remains almost universal. Many felt that the very swaddling whereby urban dwellers of the nineteenth century sought to avoid exposure to cold air actually made them hypersensitive to its effects: Samuel Sexton wrote in *Harper's* in 1879 that

> living in overheated apartments during the cold season (the temperature greatly exceeding the healthy limit of 65° to 70° Fahrenheit) develops a sensitiveness of the system, and therefore predisposes to attacks of catarrh. An excess of clothing is no less

> obnoxious than the foregoing, furs being especially dangerous in our changeable climate, as they are liable to be worn around the chest and neck in moderate weather, overheating the body, and thus increasing the liability to colds.[8]

Some saw the fear of draughts as in itself pathological: Annie Paysan Call wrote in 1894 that 'The fresh-air-instinct is abnormally developed with some of us, but only with some. The popular fear of draughts is one cause of its loss. The fear of a draught will cause a contraction, the contraction will interfere with the circulation, and a cold is the natural result'.[9] The ideal was vigorous and unimpeded circulation: the obstructions that produced stoppage of airflow and the irregularities that produced unnatural intensifications of airflow were equivalent dangers. These ideals emerge clearly in the discussions, inaugurated in 1878 by a pamphlet by John Marshall, arguing the advantages of circular hospital wards, which would prevent the formation of draughts, since 'having no blank ends like an oblong ward [it] . . . would receive light, air and wind from every direction' and 'sharp draughts across the ward, down draughts on the walls opposite and relatively near to open windows . . . would not exist.'[10]

Perhaps the strongest and longest-lasting of the magical dispositions towards the air bequeathed to us by the nineteenth century is the intense fetishism of the breath. George Catlin provoked and fanned concern about the enfeebling effects of breathing through the mouth rather than through the nose in his *Breath of Life* (1861), reissued as *Shut Your Mouth and Save Your Life* (1873). For others, breathing became a master-metaphor for the health and corruption of the body politic:

> The social lungs are so compressed – the breath of fresh knowledge is taken in with such small and feeble inspirations, and the state of the blood of the body-politic and social, is so paralytic and foul, such a poisonous sphere is given forth at every letting-out of the breath from the asthmatic, or else hectic lungs of the public pulpit, platform and press, that the social circle is impregnated with poisonous effluvia; till the whole body-politic is on the verge of a consumption of the lungs.[11]

Mesmerism, spiritualism and esoteric religion generated their own forms of respiratory discipline and promise of pneumatic bliss in parallel with these developments. Following Emmanuel Swedenborg, who described his techniques of psychic breathing, Victorian mediums and subjects of mesmeric trance claimed to be able to replace their vulgarly physiological respiration with a breathing of the spirit. At the same time, the physical operations of the breath were much in evidence in seances, with the libidinous rasping and panting of the medium entering the trance and the various ways in which breath might seem to be visibly condensed into exotic mists and plasmas. The sudden ripples of cold air betokening an otherworldly visitant (the spiritual apotheosis of the dreaded Draught), the film that flutters in the grate, the flickerings and abrupt extinctions of candle-flames by ghostly breezes, the tremulous muslin, the gusts of perfume, the whisper-music wafted on the air, the auras and aromas, the flaps and freaks and general, floating afflatus of it all, enact the kind of magical pneumatology that is parodied and bleakly diminished in Eliot's 'Gerontion':

Madame de Tornquist, in the dark room
Shifting the candles; Fräulein von Kulp
Who turned in the hall, one hand on the door. Vacant shuttles
Weave the wind. I have no ghosts,
An old man in a draughty house
Under a windy knob.[12]

The growing popularity of Eastern religions and mystical traditions among theosophists and others towards the end of the nineteenth century spread the news of the Hindu science of the five breaths, *prana*, *apana*, *samana*, *udana* and *vyana*, to the typists of Tottenham and the pale clerks of Crouch End.[13] The atmospheric science that grew up during the nineteenth century was paralleled and parodied in the complex astral topographies to be found in the writings of Madame Blavatsky and others. The motto for all this practice and doctrine might have been adapted from Leontes: 'If this be magic, let it be an art/Lawful as breathing.'[14]

Anaesthetic Revelation

Magical thinking about the air was also expressed in attitudes towards the gases of which chemical knowledge increased, from almost a standing start, over the course of the nineteenth century. Thought and matter, vision and the body, intertwined in a literal fashion in the cults that grew up around two gases in particular. One was ether, which named the universal, inconceivably attenuated medium that ran through the cosmos and transmitted light, heat, magnetism and the other forces. But the cosmic ether of the physicists had a bodily complement, in the idea of animal magnetism. The fact that ether was also the name of a volatile chemical, diethyl ether, first described in 1540 and used as an anaesthetic from 1846, when it was marketed as 'letheon' and, less officially, as an intoxicant, seemed to give this most magical, or imaginary, of substances a material embodiment, which could be felt, smelt and tasted.

Laughing gas parties began to be common in the 1820s, along with public demonstrations in theatres and playhouses.[15] In the cases both of nitrous oxide and of ether, the substances that altered the tone of the mind became alternative images of mind itself, anticipating the way in which the air in general would acquire, not just a kind of life, but a kind of intelligence, to become a thinking air. James Wilkinson saw in the breath the very union of body and mind, averring that

> [T]here is a breast in all of us where the mind and the body meet and conjoin in a compact which endures throughout life: a breast of affection between the mind of the body and the body of the mind . . . your thoughts and breaths are not only correspondent, but instantaneously coincident: if you think deeply you breathe deeply; if you hold thought you hold breath.[16]

As the century drew out, magical power passed from the air itself – traditionally the haunt of demons and carrier of malign influences – to what the air transmitted, in the secret ministry of magnetics, radio waves, x-rays, radioactivity and other kinds of unseen influence and undulation. The air seemed to become a *res cogitans*, agitated and angelically alive with transmitted thoughts and feelings, signs and messages.

Two kinds of magical thinking attached to the inhalation of nitrous oxide. First there was the belief that it extended one's powers of sympathy and understanding. For the bibber of the gas, thought was magically extended to all quarters of the universe, and users typically reported experiences of expanded, even cosmic awareness. The second effect was a powerful identification of one's thoughts with the gaseous agent of the altered state of consciousness, sometimes accompanied by a grandiose conviction that the cosmos itself had been volatilised into pure thought. As we saw in chapter Two, Humphry Davy had proclaimed during one of his N_2O benders, '*Nothing exists but thoughts! – the universe is composed of impressions, ideas, pleasures and pains*'.[17] William Drummond's poem *The Pleasures of Benevolence* (1835) lauded man's angelic powers, as they were both spread and tempered by benevolence, claiming that he

thro' ether soars
Sublime, on surer wing than that which bore
The youth of Crete, and melted in the sun:
With meteors travels, and with lightnings plays.
From gaseous worlds he draws seraphic breath . . .
Or heavenward soaring on the wings of thought,
By science plumed, beyond the comet's range,
Sweeps through infinitude.[18]

It was not until later in the century that the anaesthetic powers of nitrous oxide were recogniszed and its use became common in dentistry and surgery. But visionary adherents of the gas remained. William Ramsay lectured on his experiments with it to the Society for Psychical Research, confirming Humphry Davy's experiences in his report of 'an overwhelming mental impression that he alone was a self-centred existence on which passing events made little or no impression'. He became fully convinced of 'the truth of Bishop Berkeley's theory that all external objects are merely impressions on the mind, and have in themselves no real existence'.[19]

But the most serious and sustained claim on behalf of nitrous oxide had come twenty years earlier. In 1874 a prolific letter writer and opinionist

'Doctor Syntax and His Wife Making an Experiment in Pneumatics', coloured aquatint by Thomas Rowlandson, from *Dr Syntax in Paris; or, A Tour in Search of the Grotesque; being a Humorous delineation of the Pleasures and Miseries of the French Metropolis* (London, 1820).

on politics, metaphysics and spiritualism named Benjamin Paul Blood published a pamphlet entitled *The Anaesthetic Revelation and the Gist of Philosophy*, which was the outcome of fourteen years of reflection and experiment on laughing gas, which he had first experienced in 1860 in the dentist's chair. His pamphlet came to the attention of William James, who reviewed it attentively in the *Atlantic Monthly* and then decided to explore for himself the philosophical utility of the gas. His account of his experiences appeared as a long, unlikely appendix to an essay entitled 'On Some Hegelisms' he published in the journal *Mind* in 1882. Having soberly set out in the preceding essay his reasons for doubting the prospect of universal synthesis of opposites held out by Hegelian philosophy, James reported that the first effect of the gas was 'to make peal through me with unutterable power the conviction that Hegelism was true after all'.[20] The gas dissolved into unity all the dichotomies and contradictions on which James was wont to insist in his un-inebriated condition: 'The centre and periphery of things seem to come together. The ego and its objects, the *meum* and the *tuum*, are one' ('On Some Hegelisms', 206). James was utterly overtaken by the conviction

> that every opposition, among whatsoever things, vanishes in a higher unity in which it is based; that all contradictions, so-called, are of a common kind; that unbroken continuity is of the essence of being; and that we are literally in the midst of an *infinite*, to perceive the existence of which is the utmost we can attain. ('On Some Hegelisms', 206)

James was disappointed to find that his infinite apprehensions yielded little in the way of interesting philosophical material. Instead, his scribblings turned around the kind of magical word-play that would come to be thought characteristic of the schizophrenic, in which the decomposing of words did duty for a properly philosophical dissolution of distinctions, or, in James's terms, 'the mind went through the mere *form* of recognising sameness in identity by contrasting the same word with itself, differently emphasised, or shorn of its initial letter' ('On Some Hegelisms', 207). James treats us to some of the relics of his rapture: 'What's mistake but a kind of take? . . . Within the *ex*tensity that 'extreme' contains is contained the '*extreme*' of *in*tensity . . . Reconciliation – *e*conciliation! . . . By George, nothing but othing! . . . That sounds like nonsense, but it is pure *on*sense!' ('On Some Hegelisms', 207). The only quarter-plausible formulation found amid this fairy gold is: 'There are no differences but differences of degree between different degrees of difference and no difference' ('On Some Hegelisms', 207), which James is forced to acknowledge 'has the true hegelian ring, being in fact a regular *sich als sich auf sich selbst beziehende Negativität* [self as self-relating negativity]' ('On Some Hegelisms', 207).

The suggestion that Hegelianism will seem imperiously rational to the spaced-out gas-head seems like a gratuitous and grotesquely inconsequential swipe, especially given the skewering Hegel had endured in the essay that precedes this appendix. But there is in fact a strong continuity between the essay and its inebriate supplement, albeit one that is retroactively disclosed by the latter. Against the universal syntheses demanded and proclaimed by Hegel, James urged in the preceding essay his vision of a world of which 'the parts seem, as has been said, to be shot out of a pistol at us . . . Arbitrary, foreign, jolting, discontinuous – are the adjectives by which we are tempted to describe it' ('On Some Hegelisms', 187). Nevertheless, there are some forms of synthesis available, though these have only the form of a ground or screen that allows these jostlingly heterogeneous phenomena to subsist. James distinguishes three such screens, namely space, time and consciousness; but he devotes most of his time to the first. James's scheme requires the weakly synthesizing envelope of space in order that the discontinuous objects of the world may lie together, but apart, in what he calls 'absolute nextness' ('On Some Hegelisms', 187) or 'partaking' ('On Some Hegelisms', 191).

Hegelianism revolts against this 'partial community of partially independent powers' ('On Some Hegelisms', 191), or, its more politically-accented version, 'a republic of semi-detached consciousnesses' ('On Some Hegelisms', 204). In Hegel's system, according to James, every identity must conceal and imply a contradiction, while every contradiction in turn entails some higher, absolute unity: it is the principle of 'the contradictoriness of identity and the identity of contradictories' ('On Some Hegelisms', 195). James plays with ideas of adhesives and solvents to express his sense of Hegel's rejection of 'the sharing and partaking business he so much loathes' ('On Some Hegelisms', 194). Contradiction is 'a glue universal' – but then 'Why seek for a glue to hold things together when their very falling apart is the only glue you need?' ('On Some Hegelisms', 193). This glue is a 'universal solvent' ('On Some Hegelisms', 194), yet not that either, given that there are in fact no substantial differences to dissolve: 'To "dissolve" things in identity was the dream of earlier, cruder schools. Hegel will show that their very difference is their identity, and that in the act of detachment the detachment is undone and they fall into each other's arms' ('On Some Hegelisms', 194).

James compares Hegel's system to the Irish bull of the Kilkenny cats. The story is that a number of German soldiers stationed in Ireland amused themselves by tying together the tails of two cats and hanging them over a washing line to fight. Seeing the commanding officer approach, a trooper quickly cut the cats from their conjoined tails and they ran off. When asked the meaning of the bleeding tails, he replied that the two cats had been fighting and had eaten each other all but their tails. (Lewis Carroll seems certainly to have had the Kilkenny cats in mind in his Cheshire cat, though it is its smile rather than its tail which is the residuum). The Hegelian system, however, goes further than this:

> the Kilkenny cats of existence as it appears in the pages of Hegel are all-devouring and leave no residuum. Such is the unexampled fury of their onslaught that they get clean out of themselves and into each other, nay more, pass right through each other, and then 'return into themselves' ready for another round, as insatiate, but as inconclusive as the one that went before. ('On Some Hegelisms', 203)

Just for a moment, Hegel's mid-air dialectical somersaults suggest something of James's own characterization of the unbreachable continuity of space earlier in his essay – 'No force can in any way break, wound, or tear it. It has no joints between which you can pass your amputating knife, for it penetrates the knife and is not split' ('On Some Hegelisms', 187). But in fact, though he admits that the 'moonlit atmosphere' of Hegelian insubstantiality is contagious, such that 'the very arguments we use against him give forth strange and hollow sounds that make them seem almost as fantastic as the errors to which they are addressed' ('On Some Hegelisms', 205), James does not condemn the magical thinking or omnipotence of thought represented by the 'silly hegelian All-or-Nothing insatiateness' ('On Some Hegelisms', 204) as windily inane, as he might easily have done. On the contrary, his metaphorology sets the looseness of a world in which different things can cohabit, because they have an allowance of room – air, breathing space – against a Hegelian system that scrunches difference into a solid block of identity. This represents the triumph of thought, a true omnipotence of thought over its object, of which not a crumb would remain undigested by the ravenous operations of the dialectic.

> Bounds that we can't overpass! Data! Facts that say 'Hands off, till we are given'! Possibilities we can't control! A banquet of which we merely share! heavens, this is intolerable; such a world is no world for a philosopher to have to do with. He must have all or nothing. If the world can't be rational in my sense, in the sense of unconditional surrender, I refuse to grant that it is rational at all. It is pure incoherence, a chaos, a nulliverse, to whose haphazard sway I will not truckle. But no, this is not the world. The world is philosophy's own; a Single Block, of which, if she once get her teeth on any part, the whole shall inevitably become her prey and feed her all-devouring theoretic maw. ('On Some Hegelisms', 192)

Although, compared with the modest transactions of Anglo-Saxon empiricism, Hegelianism 'represents expansion and freedom' ('On Some Hegelisms', 186), the solidity of Hegelian totality is, for James, autistic, asphyxiating: 'In the universe of Hegel – the Absolute Block whose parts have no loose play, the pure plethora of necessary being with the oxygen of possibility all suffocated out of its lungs – there can be neither good nor bad, but one dead level of mere fate' ('On Some Hegelisms', 204). Ironically, this densely undifferentiated solidity proves also to be the terminus of James's euphoric whiffs of nitrous oxide. For, continued beyond a certain point, the universal communication afforded by the nitrous oxide turns into a kind of meaningless '*indifferentism*' ('On Some Hegelisms', 208). Suddenly, the oceanic vision shrinks to an 'intense bewilderment, with nothing particular left to be bewildered at, save the bewilderment itself. This seems to be the true *causa sui*, or "spirit become its own object"' ('On Some Hegelisms', 208). The Romantic experience of nitrous oxide had made it the means to a magisterial melting of matter into pure thought. James takes from the same 'insolence' ('On Some Hegelisms', 192) of omnipotent thought a magic that ultimately cancels and blockades itself. (Appropriate then that the recreational use of nitrous oxide should turn out to carry a risk of death through anoxia and to be a major greenhouse gas in the atmosphere.)

It is right to acknowledge that James may perhaps be accused of putting on his own kind of philosophical squeeze, in refusing to acknowledge the crucial principle of time in Hegelian thought. It is this that enables him to claim that identities are always already riven by contradictions which are already bound together in identity. It may be that the breathing space that James craves is provided for Hegel, not by space at all, but by the becoming of things in time, the principle that ventilates or opens out the closed system of the always-already.

Benjamin Paul Blood's *Anaesthetic Revelation*, which had introduced James to the links between metaphysics and nitrous oxide in 1874, had prompted similar thoughts. In his 1874 review of the pamphlet, James had identified Blood as among those who 'have helped to bring the metaphysical craving into disrepute, as being a morbid overgrowth of intellectual activity'.[21] The 'anaesthetic revelation' of Blood's title is a revelation that does not unify, but opens up a gap between the mystical experience provided by the gas and the demands of ordinary communication. So, against most pharmaceutical mysticisms, Blood 'ratiocinatively explains the gist of all philosophy to be its own insufficiency to comprehend or in any way state the All'.[22]

Blood's pamphlet may be described as an exercise in mystical immanentism. He begins from the problem of how what he calls 'the duplexity of being and thought'[23] is to be resolved: 'If we know *what* we are, still we know *as* we are, – for what is known and what knows are then the same, – and then also being and knowledge are the same, – and what we are is simply knowledge – yet knowledge *of*, or *off* what we are' (*Anaesthetic Revelation*, 6) – a play on words that has an N_2O whiff about it. Although Hegel represents the high point of the attempt to bring being and knowledge into communication with one another, it nevertheless relies upon the 'vulgar sleight' (*Anaesthetic Revelation*, 24) involved in the suggestion that the subject can coincide fully with itself as object of its own self-knowing. For it to be fully adequate to being, Hegelian thought would have to be absolute and all-encompassing, even though this would seem to contradict the principle of unfinishedness that is of the essence of being. What distinguishes Hegel is 'the determination that the logic of life shall be as life exceeding, and yet perfect as exceeding, or as including excess' (*Anaesthetic Revelation*, 19). But, for thought not only to be a complete thought of life but also fully to participate in the life of which it is the thought, it would have to have the quality of incompleteness (or be limited by its indefiniteness), and therefore to fall short as thought: 'Life is sensibly exceeding and unfinished; its logic must be exceeding and unfinished also; but so it should not be to Hegel, for logic unfinished is but diastophic, or science of the fleeting, and is ever too late for the vitality of the notion' (*Anaesthetic Revelation*, 21).

Arguing that the 'sultry Hegelian noon' (*Anaesthetic Revelation*, 29) of dialectics has taken philosophical reason as far as it can go in meeting the paradoxical demand to bring together being and knowing, Blood finally announces his 'anaesthetic revelation':

> By the Anaesthetic Revelation I mean a certain survived condition, (or uncondition,) in which is the satisfaction of philosophy by an appreciation of the genius of being, which appreciation cannot be brought out of that condition into the normal sanity of sense – cannot be formally remembered, but remains informal, forgotten until we return to it . . . Of this condition, although it may have been attained otherwise, I know only by anaesthetic agents . . . there is an invariable and reliable condition (or uncondition) ensuing about the instant of recall from anaesthetic stupor to sensible observation, or 'coming to', in which the genius of being is revealed. (*Anaesthetic Revelation*, 33–4)

The principal characteristic of the anaesthetised state of consciousness is its dissolution of all distinctions and differentiations of value, in a way that is utterly foreign to the 'formal or contrasting thought' of ordinary consciousness, or 'sanity', as Blood calls it (*Anaesthetic Revelation*, 34). The transition from the anaesthetized to the rational state 'leaves in the patient an astonishment that the awful mystery of Life is at last but a homely and a common thing, and that aside from mere formality the majestic and the absurd are of equal dignity' (*Anaesthetic Revelation*, 34). This might cast a different light on the apparent triviality of the insights recorded under the effects of nitrous oxide. In his later book *Pluriverse*, an extension of the insights and principles of the 1874 pamphlet, on which he worked for more than thirty years and was only published after his death, Blood records the effects on Oliver Wendell Holmes:

> He prepared the ether, and having placed beside his bed a small table, with pencil and paper to record his impressions on awakening, he lay down and applied the drug. Sure enough, he presently found himself just sufficiently conscious to seize the pencil, and with a sentiment of vast thought wrote something down. It proved to be these words: 'A strong scent of turpentine pervades the WHOLE.'[24]

The philosophical revelation yielded by nitrous oxide is therefore the profoundly unmagical one that there is no secret, or hidden truth, that life, in its undetermined ongoingness, is sufficient to itself: 'There is no higher, no deeper, no other, than the life in which we are founded' (*Anaesthetic Revelation*, 35). This remark seems to be echoed in James's argument, in 'On Some Hegelisms', for the irreducibility of experience to knowledge, the 'real ambiguities which forbid there being any equivalent for the happening of it all but the happening itself' ('On Some Heglisms', 190). Though it has all the apparatus and atmosphere of the mystical, Blood's vision is in fact thoroughly secular: 'The secret of Being, in short, is not in the dark immensity beyond knowledge, but at home, this side, beneath the feet, and overlooked by knowledge', James wrote in his review.[25] Blood finds in the automaticity of the breath an image for this apodictic ongoingness of life. Revelation comes not from the overcoming of the breath, that great, crazy theme of much mystical and occult writing, but from our partaking in its apartness from us:

> Every breath that we voluntarily draw is, in the cosmic sense, an irrelevant interference with divine providence. We have no need to *do* it; with or without our volition it will be done; and the determination *not* to do it – which would be in [*sic*] the least violent method of suicide – is one that nature most essentially abhors. We have no account of any man succeeding in such an attempt, although many have thus experimented.[26]

It is this acceptance of the world, and the perhaps unclosable gap between world and thought, along with the reproof that this implies to Hegel's efforts toward apnoiec omnipotence, with being inhaled utterly into thought, that seem to have appealed to William James.

James remained in correspondence with Blood for the rest of his life, his appreciation growing steadily of his philosophical and poetic writings. In the meantime, Blood seems to have drawn steadily closer to James's pluralism. Blood's pamphlet is referred to again in the 'Mysticism' chapter of James's *Varieties of Religious Experience* (1902). Here, he regrets the fact that most mystical states or systems should be so monotonously monist, with Hegel once again as its ultimate incarnation: 'What reader of Hegel can doubt that that sense of a perfected Being with all its otherness soaked up into itself, which dominates his whole philosophy, must have come from the prominence in his consciousness of mystical moods like this, in most persons kept subliminal?'[27] But by now, James has come to feel that the kind of awareness of which Blood had borne witness, though it might not be able to be rendered in philosophical terms, nevertheless needs not to be discounted by philosophers:

> [O]ur normal waking consciousness, rational consciousness as we call it, is but one special type of consciousness, whilst all about it, parted from it by the filmiest of screens, there lie potential forms of consciousness entirely different . . . No account of the universe in its totality can be final which leaves these other states of consciousness quite disregarded.[28]

In an essay on Blood that he wrote in 1910, shortly before his own death, James celebrated the fact that in Blood's work we find exemplified that optimistic anomaly, a pluralist mysticism:

> His mysticism, that may, if one likes, be understood as monistic in the earlier utterance, develops in the later ones a sort of 'left-wing' voice of defiance, and breaks into what to my ear has a radically pluralistic sound. I confess that the existence of this

> novel brand of mysticism has made my cowering mood depart. I feel now as if my own pluralism were not without the kind of support which mystical corroboration may confer.[29]

I have argued that the air is an indispensable, if also mutable, mediator in the relations between thought and world, and between thought and itself. James's engagement with anaesthetic revelation allows us to set two forms of magical thinking against one another. One is the factitiously omnipotent air of mystical monism, in which all that is solid melts into the air of thought. We might imagine a material parallel for this philosophical theme in the colonization of the air by communications, and the beginnings of the dematerialization of the world by information, as our air becomes a thinking air, and ever more the fantasized arena of our thought. The other sees in the air the principles of pluralism and discontinuity and is characterised by a thought content never to be able to catch, or consume its own tail. Through his dalliances with Blood's anaesthetic revelation, William James confronts one particular form that the omnipotence of thought took in his time. Against the airy magic of a thought that dreamed of dissolving everything into itself, James proposed a philosophy that allowed elbow room or breathing space for plurality, the pragmatism that 'means the open air and possibilities of nature, as against dogma, artificiality, and the pretence of finality in truth'.[30] We may read James's pragmatism as a salutary alternative to the Hegelianism that would be aligned to some of the most imperious forms of omnipotence of thought of the last century. Magical thinking, and the magical thinking in particular of the air, is ultimately an enactment of our perturbed, imperilled relations to our own thought, and the ever-present, and still present temptation to take our thought for the world.

four

Gasworks

Left to our own devices, we are apt to imagine the nineteenth century as an epoch of weight, mass and density, characterized by a clogging profusion of Big Stuff – factories, chimneys, bridges, girders, engines, pistons, interior furnishings and epically improbable masses of underclothing. The industrial revolution of the nineteenth century has sometimes been divided into its kinetic and sensory phases, marked by the passage from raw thermodynamic motility to the subtle modifications and extensions of sensibility effected by the electric inventions and applications of the end of the century. But the nineteenth century was itself characterized by a strange collaboration of the ponderous and the imponderable, the dense and the nebular. Britain was increasingly not just massy but gassy, and characterized not just by furnaces, turbines and sofas but also by the vapours, fumes and gases which provided the energy to make and move all this vast tonnage. This is hinted at in Emerson's essay 'Poetry and Imagination' (1876). The essay begins by evoking the dominion of common sense, to which every mind, however expansive and imaginative, must be subjected, a common sense based in the 'perception of matter':

> The common-sense which does not meddle with the absolute, but takes things at their word, – things as they appear, – believes in the existence of matter, not because we can touch it, or conceive of it, but because it agrees with ourselves, and the universe does not jest with us, but is in earnest, – is the house of health and life.[1]

But this familiar perception of matter is not definitive for Emerson, and not because of any power of the human imagination to escape the enthralments of matter, but because matter itself is volatile:

> But whilst we deal with this as finality, early hints are given that we are not to stay here; that we must be making ready to go; – a warning that this magnificent hotel and conveniency we call Nature is not final. First innuendoes, then broad hints, then smart taps, are given, suggesting that nothing stands still in nature but death; that the creation is on wheels, in transit, always passing into something else, streaming into something higher; that matter is not what it appears; – that chemistry can blow it all into gas.[2]

The material imagination of the nineteenth century had to deal increasingly with the immaterial, or with largely imaginary forms of material. The century was one in which the material and the near-immaterial entered into new and surprising alliances and exchanges, exchange becoming in the process the principal mode in which the material world was made apprehensible. Increasingly, the material world was immaterialized, with the growing dependence on invisible gases, vapours and substances, from steam (or, more strictly, water vapour), to coal gas, to the ether of space that provided such an indisputable and indispensable ground for nineteenth-century physics. At the same time, previously immaterial qualities, most particularly light and time, became increasingly subject to processes of stockpiling, investment, division, mensuration and quantification that seemed to reduce them to, or make them coextensive with, the realm of material extension. All that was solid was melting into air, to be sure: but all that had previously been air was becoming newly palpable and tractable.

One emblem of this dichotomy between the material and the immaterial, iron and gas, inertia and evaporation, is the gasholder, the huge, featureless, iron lung that was used to store the gas that for much of the nineteenth century was the principal source of fuel for light and, to a lesser extent (but one that grew more important towards the end of the century), for heating and cooking. Gasholders began to spring up all over London in the second decade of the century, growing larger and more numerous as the century wore on. By the middle of the 1820s most towns with populations greater than 10,000 had their own gasworks; this was extended to towns with populations as low as 4,000 during the 1830s and by the middle of the century only towns with populations smaller than 2,500 would not have been provided with gas, and therefore gasworks.[3] The most conspicuous feature of the gasholder was the fact that, unlike a factory building, it gave a continuous image of the amount of fuel in storage, through the movement of the drum up and down. The structure was both massive and yet governed by vacuity, to which the movements of the drum gave witness. Gasholders were the principal forms in which gasworks – the work of producing gas and the poetic work of producing a material imagination of gas – were visibly bodied forth.

The Thingliness of a Gas

Combustible gases had been known since antiquity (though they were thought of as vapours rather than 'gases'), as is attested by the survival of the suggestion of moistness in the word 'damp', which – as in 'fire-damp' and 'chokedamp' – was an early name for a gas (and survives in the German *Dampf*). John Clayton experimented with combustible gases in 1684, though his descriptions of his investigations in a letter to Robert Boyle were not published until 1739. He explains that he obtained a combustible gas by heating coal: 'At first there came over only *Phlegm*, afterwards a black *Oil*, and then likewise a *Spirit* arose, which I could noways condense, but it forced my Lute, or broke my Glasses.'[4] In 1733 James Lowther described the discovery during the digging of a coalmine of 'a vast Quantity of damp corrupted Air, which bubbled through a Quantity of Water, then spread over that part of the Pit, and made a great hissing Noise'.[5] When ignited, the gas burned with a steady flame on the surface of the water. When conducted away from the pit in a pipe, the gas continued to discharge in a stream that had remained unabated for nearly three years. Lowther described the way in which the gas might be conveyed away from its source for burning:

> The said Air being put into a Bladder . . . and tied close, may be carried away, and kept some Days, and being afterwards pressed gently thro' a small Pipe into the Flame of a Candle, will take Fire, and burn at the End of the Pipe as long as the Bladder is gently pressed to feed the Flame, and when taken from the Candle, after it is so lighted, it will continue burning 'till there is no Air left in the Bladder to supply the Flame.[6]

It was not until the 1780s that the possibility of collecting and transporting combustile gas and burning it for light or heat began seriously to be considered. Lord Dundonald used a by-product from a process of distilling tar from coal during the 1780s in Culross Abbey, though he brought the gas up to the abbey in tanks, rather than piping it.[7] In 1792 William Murdoch succeeded in lighting his house in Cornwall with coal gas, and subsequently used it to light Boulton and Watt's factory in Soho in 1798. But the establishment of gas lighting would depend upon the arrival on the scene of an extravagant, obscure impresario, in the person of Friedrich Albert Winzer, or after his arrival in England, Winsor. Winsor, who had learned of the possibilities of coal gas from the *thermolampe* invented by the French chemist Philippe Lebon, with whom he worked for a time, and who was an amazing and, for many, infuriating compound of crackpot and smart alec. It was he who seems first of all to have recognized, not only that gas was useful, but also that, in order to be put to use, it would need to be considered not as a transportable material, like coal, but as a public utility, which would require a massive physical and political infrastructure to allow it to be integrated into the social fabric, which it would in the process transform. It was Winsor, in short, who saw that gas would need to be constructed as a *mediate material*. Following in the tradition of a Dutch chemist called Diller, who had used some kind of combustible gas (possibly hydrogen) to mount a display of what he called 'Philosophical Fireworks' at the Lyceum Theatre in 1788, Winsor produced a series of exhibitions and lectures aimed at publicizing the properties of coal gas. By 1807, following a successful demonstration of gas lighting on the walls of Carlton House Garden, Winsor had issued a prospectus for what he called a 'National Light and Heat Company', which he intended to be in effect a nationally sponsored private monopoly.

Winsor made much of the fact that purifying coal into a series of useful products would improve the terrible pollution caused by the burning of coal wholesale:

> LONDON, the emporium of the commerce of the world has been, and may justly be stigmatised, the *Emporium of Smoke*! because there is not a city or country in the universe, half so much soiled and obscured by the offensive vapours of pit coal, issuing, day and night, from the innumerable manufactories and chimnies in the realms of Great Britain, where, alas, the immense clouds of smoke and darkness, generate showers of soot, and obscure that beautiful solar Light which the Almighty intended for our comfort and happiness.[8]

He also estimated that the net profit to the nation from the collection and use of combustible coal gas, including saving of wasted resource and tax revenue, would be £111,845,294.[9] This utilitarian argument may have made a strong appeal to the investors from whom he sought the massive sum of a million pounds in start-up capital. But, though he saw the big picture with regard to gas more clearly than others had, Winsor's grasp of technical detail was alarmingly tenuous. In a question-and-answer pamphlet intended to rebut objections to his proposal, Winsor tried to reassure his readers about the safety of coal gas. Pure coal gas would never explode, he says, but simply burn (this is broadly true). But then he goes a step too far:

> *Q*. . . . suppose a room full of gas, and you enter it with a candle.
> *A*. It will *never* inflame, because it is intermixt with the air of the room.[10]

Winsor had also reported on the fertilizing effects of gas in a pamphlet entitled *Analogy Between Animal and Vegetable Life*, which reported that:

G. M. Woodward, engraved by Thomas Rowlandson, *A Peep at the Gas Lights in Pall-Mall* (1809). Various characters pass comment on the new gaslights: a knowledgeable gentleman, a rustic, an Irishman ('Arragh honey, if this man bring fire thro water we shall soon have the Thames and the Liffey burnt down – and all the pretty little Herrings and Whales burnt to cinders'), a Quaker ('Aye, Friend, but it is all Vanity: what is this to the inward light?'), and a lady of easy virtue, with her footpad friend ('If this light is not put a stop to – we must give up our business. We may as well shut up shop').

> The result and conclusion of all my arguments and observations are, that the *application of Gas Lights to Hot and Green-houses*, must be very beneficial; as by the operation of the *Patent Light Stove*, the principal nourishment for vegetation, such as *Light* and *Caloric, Hydrogen, Nitrogen, Carbonic Acid Gas*, &c. are plentifully generated, and may be applied without interruption, in a much greater proportion than even the most powerful action of solar rays can possibly produce, in the common operations of nature.[11]

Winsor's proposal met with considerable and concerted opposition. One objector declared that 'no one would rely upon so foolish and unlucky a thing as gas'.[12] Winsor's emphasis on the purity of gas was mocked in a poetic squib issued in the year following his first prospectus. Feigning to reject the views of those who 'say thy Gas imparts a fetid gust / To nasal nerves, and gives to metals rust', the poem sardonically pumps up Winsor's rhetoric of purity into a philosophical principle:

> O! could thy Gas with equal power convey,
> To the mind's eye an intellectual ray,
> With flame ætherial decomposed from coal,
> Illumine and HYDRO-CARBONATE the soul;
> Our streets, so long with walking Idiot's curst,
> Where Dunce the second elbows Dunce the first,
> Might see parade their crowded path along,
> A novel species – an enlightened throng [. . .]
> The active principle thy Gas supplies,
> Haply may bid a new creation rise;
> With atmospheric oxygen combin'd,
> Bid human nature leave its dross behind;
> Like the wise serpent cast away its case,
> But like an oxyde still retain its base;
> And Man, with faculties *sublim'd and rare*,
> Forsake his humbler walk, and tread on air.[13]

Winsor's great gas wheeze inevitably provoked comparisons with the South Sea Bubble. Lady Bessborough wrote in 1807 of the current excitement about the investment opportunity:

> What can occasion such a ferment in every house, in every street, in every shop, in every Garrett about London? It is the Light and Heat Company. It is Mr. Winsor, and his Lecture, and his gas, and his patent, and his shares – those famous shares which are to make the fortune of all of us who hold them, and probably will involve half England in ruin, me among the rest, and prove a second South Sea Scheme.[14]

One objector to his first project declared that 'Mr Winsor's *puffs and projects have at last failed*; and no one can be found to believe in the sanguine calculations of twelve thousand per cent. profit per annum.'[15] But these mockeries in fact help us to recapture the association between Winsor's sociopolitical utopianism and the more poetic and philosophical elation encouraged by the remarkable discoveries made in the chemistry of gases at the end of the eighteenth century. Winsor's inflammable air would have been easy to associate with the exhilarating gases investigated by Priestley, Lavoisier and Davy, which promised ecstatic visions in the worlds of art and poetry. The association is made by the reprinting as a frontispiece to the *Heroic Epistle* of James Gillray's satirical portrait of Humphrey Davy's demonstrations of nitrous oxide and other gases in his lectures at the Royal Institution in 1802. More even than this, the *Heroic Epistle* sardonically identifies what would become an important principle of gas, namely that it was a commodity based upon a new, ambivalent kind of matter, or body, that in its turn entered into composition with new kinds of social body:

Thomas Young experimenting on Sir J. C. Hippisley at the Royal Institution, London, with Humphry Davy holding a bellows behind them; James Gillray, *Scientific Researches! – New Discoveries in* PNEUMATICKS! *– or – an Experimental Lecture on the Powers of Air*, hand-coloured etching (London, 1802).

> Bid Gas and Coke in statute pride display
> Their patent blaze by night, their heat by day,
> And give to airy nothings (LIGHT and FLAME),
> A BODY CORPORATE in deed and name.[16]

Distance

Gaslight appears in the fiction of the nineteenth century increasingly as an evoker of visual atmosphere, or ophthalmic accessory. As Bruno Latour has suggested, it is perhaps the very tendency of objects in such wide use to sink into the background that makes it imperative to find ways of restoring them to articulation:

> Objects, by the very nature of their connections with humans, quickly shift from being mediators to being intermediaries, counting for one or nothing, no matter how internally complicated they might be. That is why specific tricks have to be invented to make them talk, that is, to offer descriptions of themselves, to produce scripts of what they are making others – humans or non-humans – do.[17]

Wolfgang Schivelbusch has suggested that the most important principle of gas lighting, which was reinforced rather than reversed by electric lighting, was that of distance. Not only was the gas flame much less intimate than the candle flame, the fuel was not produced locally, but at a distance. For Gaston Bachelard, whose meditation on candles and lamps helps to define the world that gaslight both supersedes and discloses, the flame is ‘a becoming-being, a being-becoming’, in which matter strives visibly into the form of light.[18] The gas jet abstracts even the idea of the flame, through the use of magnifying and diffusing shades and frosted glass to diffuse and regularize the flame into panes of light.[19] The aim was to produce uniform intensity in place of the fluctuations and irregularities typical of extended matter. What had previously been proximate, iterative and particular was to become remote, absolute and general. This kind of abstraction partook more of an idea or principle than of the conditions of matter. Matter was to be rationalized as well as theorized, made to approach more and more closely to the principles of its operation. Gas represents a kind of purification of things into the abstract matter of which they are composed. Matter thereby becomes both less and more 'material'.

The development of the gas meter, which was among the most important and necessary technical developments in the introduction of gas, made gas an abstract material. Even the unit in which it was reckoned and purchased, the cubic foot, made little empirical sense to customers. The introduction of the coin-in-the-slot meter in the late 1880s converted the unit to a penny's worth. Gas announces the arrival of what Bachelard mournfully calls 'an age of administered light'.[20] Bachelard suggests that every flame is a kind of rising striving: 'one could say that everything upright, everything vertical in the cosmos, is a flame'.[21] Even before the development of the incandescent gas-mantle, one of the earliest signs of the denaturing of the flame in gaslight was the demonstration that flames could be made to burn at any angle, including straight downwards. One of its early enthusiasts wrote that the properties of gas 'render it particularly fit for ornamental illumination. As there is nothing to spill, the flame may be directed either downwards, upwards, or horizontally; and the points from which it issues may be disposed in any form that taste or fancy may suggest'[22] The fact that, unlike candles or even electric lights, both of which were portable to some degree, and could shine upwards, gaslights were nearly always positioned so that they cast their light downwards may also explain their lowering capacity.

The most important form of distancing or abstraction was the separation of light from combustible matter. Jonathan Crary has suggested that the 1820s saw the situating of vision within the body as a physiological event. If this is true, then the artificial light supplied by gas seemed like a kind of autonomous vision, a harsh provocation to and denial of the embodied eye. Gaslight seems to have been experienced as a kind of flattening or banishing of more comfortable and familiar kinds of gradations. The commonest adjective to apply to gas seems to have been 'garish'. The word derives perhaps from 'gaure', to stare, suggesting that at which one is constrained to stare, as well as, perhaps, to be reflexively stared at by. As Bachelard notes, the psychological principle governing light sources is that that by which we see, itself sees us: 'everything that casts a light, sees'.[23]

Gas was associated not just with crepuscular gloom or flicker, but also with its opposite – a gaunt flattening. The associations of gas with theatricality, theatres being early adopters of gas, spread across into non-theatrical locations – especially shops and streets – to turn them into places of display and exhibition; as Lynda Nead has said '[g]aslight turned the London streets into a stage'.[24] The gaslit theatre of the streets reduced everything in them to the fact of their being seen, a kind of penury of shownness. In her account of the theatrical life, Olive Logan described going on stage as 'a leap in the dark, but a leap into the light – into the gaslight, the streaming, gleaming, all-revealing gaslight'.[25] Later, she quotes another judgement on the exposure of the performer:

> 'Men and women who are compelled by their vocation to move before the world in a perpetual glare of gaslight, and to submit to a surveillance which is ceaseless, and to a judgment which is seldom charitable, are sure to be suspected however innocent, and equally sure to be detected however cautious'.[26]

Where the flames of candles and lamps seemed to suggest an 'igneous time'[27] of slow, rhythmic accumulation and elapse, gaslight could be lit and extinguished instantly, without prelude or residue. Even before the arrival of electric light, which did not require one to be in the vicinity of the light itself to turn it on and off, gas seemed to offer a new experience of rapid alternation between opposite states, effected by the switch:

> A finger on the switch is enough to make dark space immediately bright. The same mechanical gesture causes the inverse transformation. A little click says *yes* and *no* with the same voice. Thus the phenomenologist has the means of placing us alternately in two worlds, which is as much as to say 'two consciousnesses.' With an electric switch one can play the games of *yes* and *no* endlessly. But in accepting this mechanism the phenomenologist has lost the phenomenological density of his act.[28]

Gaslight also suggested a kind of flashbulb vision, substituting sudden flarings, irradiations or exposures for the more restrained, approximate, ambivalent reciprocities of light, eye and object. Gas provided sudden revelations and reversals of perception, abrupt transformations: 'Oliver felt her hand tremble; and, looking up in her face as they passed a gas-lamp, saw that it had turned a deadly white.'[29] In Wilkie Collins's *Basil*, an obsessed pursuer suddenly sees himself through the intermediary of a gas-lamp:

> I remember passing two men in this way, in some great thoroughfare. They both stopped, turned, and walked a few steps after me. One laughed at me, as a drunkard. The other, in serious tones, told him to be silent; for I was not drunk, but mad – he had seen my face as I passed under a gas-lamp, and he knew that I was mad.
>
> 'MAD!' – that word, as I heard it, rang after me like a voice of judgment.[30]

Though gaslight is nowadays associated with crime, mystery and enigma ('[t]his gaslight is a sort of an artificial moonlight, you know, and sometimes it produces the same effects'[31]), its raw, uncompromising glare gave it the reputation of dispelling romance and spectrality. Worrying about being haunted by the ghost of the murdered George Talboys, Robert Audley in Braddon's *Lady Audley's Secret* looks to gas for reassurance:

> 'I haven't read Alexander Dumas and Wilkie Collins for nothing,' he muttered. 'I'm up to their tricks, sneaking in at doors behind a fellow's back, and flattening their white faces against window-panes, and making themselves all eyes in the twilight. It's a strange thing that your generous-hearted fellow, who never did a shabby thing in his life, is capable of any meanness the moment he becomes a ghost. I'll have the gas laid on to-morrow and engage Mrs Maloney's eldest son to sleep under the letter-box in the lobby. The youth plays popular melodies upon a piece of tissue paper and a small-tooth comb, and will be quite pleasant company.'[32]

Gas seems to emaciate what it illuminates, leaving no excess beyond what is rendered up to sight. Graham's first sight of Julie Caumartin in Bulwer-Lytton's *The Parisians* suggests this hollowing effect:

> On first entering the gardens, Graham's eye was attracted and dazzled by a brilliant form. It was standing under a festoon of flowers extended from tree to tree, and a gas jet opposite shone full upon the face, – the face of a girl in all the freshness of youth. If the freshness owed anything to art, the art was so well disguised that it seemed nature. The beauty of the countenance was Hebe-like, joyous, and radiant, and yet one could not look at the girl without a sentiment of deep mournfulness. She was surrounded by a group of young men, and the ring of her laugh jarred upon Graham's ear.[33]

Graham's friend Frédéric explains that Julie's appearance is indeed the product of an eclipse: "'A little while ago her equipage was the most admired in the Bois, and great ladies condescended to copy her dress or her *coiffure*. But she has lost her splendour, and dismissed the rich admirer who supplied the fuel for its blaze, since she fell in love with Gustave Rameau.'"[34]

The powers of gaslight not just to exhibit, but cruelly to amplify the effects of dissipation are on display in Emerson Bennett's *Ellen Norbury*:

> [T]he stately Mrs Pinchbeck sailed in, dressed in silk. She was a large woman, and, in her younger days, had probably been good looking; but her features now were coarse, and a near inspection showed the crow's-feet around her small, keen eyes, notwithstanding a pretty free use of cosmetics. Her hair and teeth were false, her eyebrows pencilled, and her large flabby cheeks painted. Still, by gas-light, she was as passably comely as could be expected of a woman verging on fifty, who had spent a good portion of her time in fashionable dissipation.[35]

Perhaps the most memorable distinction between the inhumanly enhanced mode of vision provided by gas and embodied human vision is provided in Sam Weller's protest, in response to Serjeant Buzfuz's question '"Have you a pair of eyes, Mr Weller?"': '"If they was a pair o' patent double million magnifyin' gas microscopes of hextra power, p'raps I might be able to see through a flight o' stairs and a deal door, but bein' only eyes, you see, my wision's limited"'.[36] Perhaps not many readers would have known that since 1824, gas microscopes had indeed been available for the enhancement of vision. Julian Wolfreys sees this passage as one of the ways in which the writing of vision in *Pickwick Papers* evokes 'the ghostly *revenant* of technology-to-come'.[37] In a note, he acknowledges the contemporary existence of the gas microscope, but somewhat petulantly protests that 'literature . . . is irreducible to fact, date, or event . . . Dickens should perhaps be read with an eye directed more to the spirit, rather than the letter of the historical'.[38] In fact, it need not come to a choice, or to precisely this kind of choice, between historical fact and future projection. For the device in question used gas merely as a source of illumination, which gave the clarity of image that would allow projection on to a screen (thus, a gas microscope may have been part of the kind of magic lantern capable of throwing 'the nerves in patterns on the screen' that Prufrock imagines a century later). Andrew Pritchard dates the first public use of such a device to 1824, when it was employed by George Birkbeck in two lectures on optical instruments delivered at the London Mechanics' Institution, in one of which

> he took occasion to delineate on a screen, by means of a large magic lantern, representations of magnified objects intensely illuminated by the light emitted during the combustion of lime by hydrogen and oxygen gases, and to indicate the practicability of applying successfully this method of illumination to the microscope.[39]

But the rhetoric of gas might have encouraged readers in 1837 (and Sam Weller himself) to have imagined for gas, not just the power to illuminate things more clearly, but also to see into or even through them. The gas is both a lens and an x-ray before the fact, and Weller's exasperated exaggeration does indeed seem to peer into the future of vision. George Eliot hints, more soberly, at a similar association, in her account of the influence of the biologist Bichat on Lydgate in *Middlemarch*:

> No man, one sees, can understand and estimate the entire structure or its parts – what are its frailties and what its repairs without knowing the nature of the materials. And the conception wrought out by Bichat, with his detailed study of the different tissues, acted necessarily on medical questions as the turning of gas-light would act on a dim, oil-lit street, showing new connections and hitherto hidden facts of structure which must be taken into account in considering the symptoms of maladies and the action of medicaments.[40]

Here, the x-ray spectacle of gas does not merely dissolve the material fabric of things, but rather shows the network of connections that make the body as a whole intelligible, just like the underground reticulations of pipework that made gas supply possible.

Sightly Operations

If consumers of gas were separated in time and space from gas's point and process of production, it was by no means entirely inaccessible to them. For one thing, gasworks tended to be built in the heart of cities, initially to save on the costs of piping and subsequently through a kind of inertia. The smell and sight of the gasworks, with its accompanying gasholder, were a reminder of the ugliness of the process whereby the city's purified light was obtained. Nor were the insides of gasworks inaccessible, for accounts of the demanding conditions of gasworks, like this one from 1865, appeared regularly:

> In the hottest of the work men frequently strip to the waist, and many will, while still reeking with perspiration, stand under the open louvres in the roof, no matter how strong the draught. At many of the works, the men are provided with an unlimited supply of a drink composed of oatmeal and water, 'skilly,' and of which the carbonisers at the Chartered Gas-works (Horseferry-road) are found to consume an average of seven quarts a day, so great is the thirst occasioned by their work. One of the men at these works while incautiously drinking freely, not long since, of cold water when overheated, dropped dead to the ground.[41]

Visitors to London, like Flora Tristan, were struck by the infernal grandeur of the gasworks and the workers who sustained them. Her 1840 account of the Horseferry Road gasworks emphasizes the physical horror of the place:

> [I]t was *hot*. I could not remain in this inferno; the air was unbreathable, the gas fumes were making me giddy, the heat was suffocating . . . At every moment one is assailed by noxious fumes. I came out from the shed I was under, hoping to breathe purer air in the yard, but I was pursued everywhere by the foul exhalations of gas and the smell of coal, tar, etc.[42]

And yet she is also moved to appalled raptures by the spectacle of the red-hot coke being removed from the retorts:

> Nothing could be more terrible, more majestic than these mouths vomiting flames! Nothing could be more magical than the cellar suddenly lit up by the burning coals which were rushing down like a waterfall cascading from the heights, and similarly engulfed in the abyss! Nothing could be more terrifying than the sight of the stokers, streaming with perspiration as if they had just come out of the water, and illuminated from all around by those dreadful sheets of fire whose tongues of flame seem to be darting out to devour them. Ah no, it would be impossible to see a more awesome spectacle![43]

Gustave Doré's engraving of the Lambeth gasworks in his *London Pilgrimage* (1872, with Blanchard Jerrold) offers a slightly more oblique perspective, but one that is intended to suggest similarly infernal associations. The point of the picture seems to be the rather strange contrast between the huddled, haggard misery of the figures gathered listlessly in the foreground and the desperate industry of the stokers glimpsed in the retort room in the background. It is not clear what transaction is taking place over the person of the obligatory swaddled mite on the right of the picture, but it certainly seems an insalubrious atmosphere for one of such tender years to be imbib-

Gustave Doré, 'Lambeth Gasworks', from Gustave Doré and Blanchard Jerrold, *London: A Pilgrimage* (London, 1872).

ing. Doré seems to be showing in the background the same stage in the process as Tristan, the clearing out of the retorts, using the long scoops of which an example is being held by the central figure in the foreground, in a bizarre and unintelligible parody of a sentry on guard. The stokers behind seem rather to be engaged in a desperate struggle against the advance of the flames than about to remove the products of the distillation.

We are offered here a kind of primal scene of the production of gas, the apparent concentration, in one visible place and time, of the toil and suffering that are required to wrest light from the sediments of the ages. We might recall that two of the most important pressures that prompted the development of gas as a public utility, after the long dormancy of more than a century during which the viability of using it to provide lighting had been well known, were the demands of new, large public spaces, especially factories and places of entertainment.[44] The factory-theatre is a predictable hybrid outcome from this demand.

But there is no simple overcoming of alienation or bringing to light of industrial process, in Doré's illustration. The clearest, most legible part of the illustration is the foreground, in which materiality, in the lumps of coal visible in the carriers, is merely matter, and the exhausted workers appear just as inert and purposeless. Strenuous action is visible in the background of the picture, where the coordination of the gangs is in contrast with the arbitrary aggregations of the figures in the foreground, but their action seems to have thinned them to spectres. The real process of producing the 'wild spirit' of the gas remains invisible, occurring somewhere between the lines of retorts and the large gas-pipe leading away in which the top right of the picture to the gas-holder. There is no primal scene in which the matter of gas, the source of light, lets itself be seen. As Dickens remarked, in his description of the gas-making process in his article 'The Genii of the Lamps', which appeared in *Household Words* in October 1861, '[t]he manufacture of gas, although it includes many beautiful scientific processes, is not on the whole, a sightly operation. What is not seen may be refined and interesting; but what is seen decidedly savours of pandemonium'.[45]

A Part of Ourselves

Inevitably, though more slowly and unevenly than we might perhaps expect, gas combustion also proved to be resistant to the process of abstraction. The sensory paradox of gas is that a substance that must at all costs and at all stages of its production and distribution remain safely enclosed and insulated should have effects that spill beyond those enclosures. Increasingly, the light of the gas began to be thought of as itself a kind of gaseous emanation, which entered into competition and composition with the air. Although early promoters of gas emphasized the fact that the fuel provided both light and heat, the great technical problem for later designers of gas appliances and installations was how to maximize the lighting output of gas jets while reducing to a minimum the tendency to heat the air. One manufacturer of ventilation systems for gas lighting warned that:

> Like fire and water, gas is a good servant but a bad master; and in ninety-nine houses out of a hundred it is simply a licensed despot, who revenges himself for the light he cannot help giving, by poisoning the very air to which he is indebted for life and brilliancy.[46]

However assiduously engineers and designers tried to isolate the gas burner from its environment, it proved to be closely dependent upon and determining of it. Gas burned differently depending on variations in atmospheric pressure, a rise of an inch in barometric pressure increasing the luminosity of the flame by 5 per cent. Humidity also affected the flame, as did the pressure of the gas (too much pressure produced over-aeration and a reduction in the intensity of the light).[47]

Gas was therefore essentially *fluctuant*, in pressure and quality. The residual feature of gas was the variability of its quality, or the quality of its variability, against which designers and advertisers struggled. Lynda Nead suggests that the most important effect of gas was the way in which it created fascinating, seductive, and also sinister urban *chiaroscuro*.[48] George Cukor's film *Gaslight* (1944) emphasises the flickering of the gaslights, for this is the way in which a husband convinces his wife that she is losing her mind. As a result of this, the term 'to gaslight' means to disorientate. In the end, the continuing success of gas has come to consist in this very aptness to fluctuation, the 'organic' sensitivity of gas that makes it more controllable and responsive than electricity in cooking.

Gas had implications for physiology as well as physics. The early nineteenth century inherited from the heady days of gas discovery in the 1780s a strong, and seemingly inextinguishable, predisposition to believe in the health-giving effects of various kinds of gas. Not only were the tonic powers of oxygen and nitrous oxide touted, but also other, more surprising gases, like carbonic acid gas, were found to have therapeutic or vitalizing powers. Most surprisingly of all, given the widespread knowledge of how toxic it was, coal-gas was included in this roster of medicinal gases. F. A. Winsor made some very enthusiastic claims, during his vigorous campaign of publicity on behalf of gas lighting and heating, about the health-giving properties of coal gas. One of them is in answer to the imagined question of an objector:

> Q. But is it not hurtful to respiration?
> A. Not in the least! – On the contrary, it is more congenial to our lungs than vital air, which proves too strong a medicine, because it only exists from one-fifth to one-fourth in the atmosphere; whereas, inflammable air exists above two-thirds in the animal and vegetable kingdoms, in all our drink and victuals. It forms a part of ourselves.[49]

We learn from the *Heroic Epistle to Mr Winsor* that he went even further in his claims for the health-giving properties of his gas: 'Mr Winsor, in his Lectures, states that he has cured himself of a constitutional asthma, by superintending the works at his stoves, and inhaling his own hydro-carbonic gas, and invites any persons afflicted with disorders of the lungs to attend at his house, and try the remedy gratis'.[50] One historian of the gas industry wrote in 1988 of his own 'childhood memories of being dragged to the Torquay and Paignton Gas Company where the smell of the sulphurous discharge was supposed to induce sickness and cure whooping cough'.[51]

In many of the salacious tours of urban night life in which gaslight featured the toxicity of gas blended with its intoxicating accompaniments, as, for example, in George Ellington's complaint in his *Women of New York* that 'crowds of boys and girls of a susceptible age meet together under the intoxicating influence of music, gaslight, full dress, late suppers, wines and punches'.[52] Dickens associates gas with the effects of intoxication, too, in the account of Mr Pickwick's fluctuant process of inebriation:

> The wine which had exerted its somniferous influence over Mr Snodgrass, and Mr Winkle, had stolen upon the senses of Mr Pickwick. That gentleman had gradually passed through the various stages which precede the lethargy produced by dinner, and its consequences. He had undergone the ordinary transitions from the height of conviviality, to the depth of misery, and from the depth of misery, to the height of conviviality. Like a gas lamp in the street, with the wind in the pipe, he had exhibited for a moment an unnatural brilliancy: then sunk so low as to be scarcely discernible: after a short interval, he had burst out again, to enlighten for a moment, then flickered with an uncertain, staggering sort of light, and then gone out altogether. His head was sunk upon his bosom; and perpetual snoring, with a partial choke, occasionally, were the only audible indications of the great man's presence.[53]

Gas gradually fell under suspicion as the century wore on, as domestic gas became associated with sewer-gas and the many other noxious vapours with which the miasmatist Victorian hygienists were much preoccupied. Gasworks were accused, often with good warrant, of polluting the air, water and ground in their vicinity, ironically, often as a result of the by-products of the processes employed to purify the gas.[54] In France, Evariste Bertulus blamed gas pipes for the vitiation of wells and water supplies and even blamed it for epidemics of yellow fever and typhus in Marseilles.[55]

The adoption of gas for cooking towards the end of the century, and the spread of gas ovens into domestic houses, especially of the working class, made for a new kind of bodily intimacy with gas. Indeed, one may say that, with the arrival of electricity, gas came to seem more familiar and companionable, for all its apparent inconvenience, and perhaps partly because of it. Where electricity meant excitement, gas meant comfort:

> People little reflect, how much of comfort they owe to gas. The cheapness of the light, the ease with which it is manipulated, its handiness and homeliness, so to speak – because the gas is always there, ready at the moment when wanted; its cleanliness, its safety, are all advocates of gas lighting, and speak eloquently in its favour. Gas is like a good and willing and trustworthy servant. It is not obtrusive or despotic in its manifestations, as is the electric light, nor dirty and slatternly like candles, oil, and the oil lamp. These latter oppress the mind, because you are never sure of them, or rather you are always sure of their uncertainty: but gas, like a good stomach and liver, goes cheerfully and brightly on, minding its own proper business.[56]

A somewhat surprising confirmation of the bodily intimacy of gas is the adoption and growing popularity of gas poisoning as a method of suicide. Histories and surveys of suicide do not identify the head-in-the-oven method of suicide as becoming common until the early years of the twentieth century. The *Lancet* reported in 1908 that a coroner in Manchester had requested that newspapers not publish details of the method employed by one gas suicide, in order to discourage copycatting, suggesting that the technique may have seemed quite novel.

> A verdict of suicide during insanity was returned yesterday, in the case of a clerk who had lost his situation through drink and poisoned himself with gas. The city coroner pointed out that suicides through gas poisoning were on the increase. Three had occurred within a week and 'he could not help thinking that some sort of suggestion was going on.' Whether due to harmful literature, describing how it could be accomplished by gas, or whether it was through details being given in the daily press, he did not know; but 'he thought he was justified in asking the press on this occasion to keep out the details of the case.' He thought if this were done the 'epidemic' would probably cease. There is no question as to the wisdom of this suggestion and in this instance the local papers seem very properly to have acted up to it. [57]

The historian of suicide Olive Anderson suggests that this may have been the case in Manchester in particular owing to the slow penetration of gas into working-class homes there.[58]

Mediation

Gas does not represent the simple thinning or abstraction of matter. Rather, gas procures and represents what may be called *mediate matter*. George Sala took a hint from the opening of *Bleak House* for his literary gazetteer of the capital, *Gaslight* and *Daylight in London*, making gas a kind of universal witness and conductor:

> An impartial gas, it shines as brightly on the grenadier's quart-pot as on the queenly crown. A convivial gas, it blazes cheerfully in the mess room of the Beauchamp Tower. A secretive gas, it knows that beneath the curtains and flags of that same mess-room there are dark words and inscriptions cut into the aged wall – the records of agony and hopeless captivity, anagrams of pain, emblems of sorrow and hopes fled and youth and joy departed.[59]

The gas becomes the image and the medium of Sala's own writerly perambulations, both bringing light and the whispering of secrets. The most important feature of the gas is that it goes everywhere, dissolving and pervading, making everything accessible even as it inhabits privacy. Gas has two modes of propagation: as illumination it exposes and integrates, dispelling secrecy and enigma; as diffusive matter it infiltrates and propagates, itself secret and invisible. Because chemically and scopically it is everywhere the same, it can adapt itself to any circumstances and insinuate itself into every space:

> As I walk about the streets by night, endless and always suggestive intercommunings take place between me and the trusty, silent, ever-watchful gas, whose secrets I know. In broad long streets where the vista of lamps stretches far far away into almost endless perspective; in courts and alleys, dark by day but lighted up at night by this incorruptible tell-tale; on the bridges; in the deserted parks; on

> wharfs and quays; in dreary suburban roads; in the halls of public buildings; in the windows of late-hour-keeping houses and offices, there is my gas – bright, silent, and secret. Gas to teach me; gas to counsel me; gas to guide my footsteps, not over London flags, but through the crooked ways of unseen life and death, of the doings of the great Unknown, of the cries of the great Unheard. He who will bend himself to listen to, and avail himself, of these secrets of the gas, may walk through London streets proud in the consciousness of being an Inspector – in the great police force of philosophy – and of carrying a perpetual bull's-eye in his belt.[60]

Chemically purified, gas is epistemologically compounded, and imbricated. The gas network, more complex and far-reaching than the water network, produced a chain of collaborations and corroborations. The most extended enactment of the mediating functions of this material is Dickens's *Bleak House*. In the opening sequence of the novel, gas forms part of the diffusive series of elements – mud, fog, gas – which are connected by their shared capacity for interfusion. Amid the November fog, the gas is not a source of illumination, but rather itself a visible element, as though the gas were shining through itself:

> Gas looming through the fog in divers places in the streets, much as the sun may, from the spongey fields, be seen to loom by husbandman and ploughboy. Most of the shops lighted two hours before their time – as the gas seems to know, for it has a haggard and unwilling look.[61]

Gas belongs to the new world, rather than the flickering world of the Dedlocks, as is suggested by the contrast between gas and oil lamps during Sir Leicester Dedlock's slow dissolution:

> The day is now beginning to decline. The mist, and the sleet into which the snow has all resolved itself, are darker, and the blaze begins to tell more vividly upon the room walls and furniture. The gloom augments; the bright gas springs up in the streets; and the pertinacious oil lamps which yet hold their ground there, with

> their source of life half frozen and half thawed, twinkle gaspingly, like fiery fish out of water – as they are.[62]

Gas belongs to the explosive economy of the novel, as suggested by Phil's roster of accidents, in which matter and flesh promiscuously intermingle:

> [W]hat with blowing the fire with my mouth when I was young, and spileing my complexion, and singeing my hair off, and swallering the smoke; and what with being nat'rally unfort'nate in the way of running against hot metal, and marking myself by sich means; and what with having turn-ups with the tinker as I got older, almost whenever he was too far gone in drink – which was almost always – my beauty was queer, wery queer, even at that time. As to since; what with a dozen years in a dark forge, where the men was given to larking; and what with being scorched in a accident at a gasworks; and what with being blowed out of winder, case-filling at the firework business; I am ugly enough to be made a show on![63]

But the explosiveness of *Bleak House* is matched by the tendency of matter to aggregate and congeal. The distinguishing feature of a gas, as explained in a popular exposition of its chemistry in the *Edinburgh Review* for 1809, was that it was 'an invisible elastic fluid . . . which no cold nor affusion of water can condense or absorb'.[64]

But Dickens's gas tends to precipitate or deposit itself in reluctant sludge. The winds are the 'messengers' that convey 'Tom's corrupted blood', but that corruption is imagined as an impossible kind of air-borne slime, both volatile and viscous.

> There is not an atom of Tom's slime, not a cubic inch of any pestilential gas in which he lives, not one obscenity or degradation about him, not an ignorance, not a wickedness, not a brutality of his committing, but shall work its retribution, through every order of society, up to the proudest of the proud, and to the highest of the high.[65]

Dickens's gas is no 'incorruptible tell-tale',[66] for adulteration is the very essence and method of its tale-telling. The two sides of gas, its power of illumination and its tendency to pollute, cross over curiously in the description of the churchyard where Nemo is buried. Dickens calls for light, not in order to dispel the gloom of the place, but in order to keep it at bay:

> Come night, come darkness, for you cannot come too soon, or stay too long, by such a place as this! Come, straggling lights into the windows of the ugly houses; and you who do iniquity therein, do it at least with this dread scene shut out!

The gaslight burning above the iron gate seems to provide only intermittent illumination, fluctuating grudgingly between the immaterial condition of light and the cloying condition of grease.

> Come, flame of gas, burning so sullenly above the iron gate, on which the poisoned air deposits its witch-ointment slimy to the touch! It is well that you should call to every passer-by, 'Look here!'.[67]

The passage between the airy and the slimy is suggested neatly in that 'witch-ointment', a reference to the medieval and early modern belief that witches greased their brooms with 'flying ointment' brewed up from toxic herbs and the fat of boiled babies to assist their flight. Here the gas both illuminates and leaks into the 'poisoned air' condensing in the form of slime.

Rather than being a raw material, which the network simply supplied, gas was an ambivalent material – a modern *spiritus silvestris*, or wild spirit, that itself produced and transposed positions and values. It is, as Bruno Latour has suggested, not just a neutral intermediary, that connects things up and passes unchanged and unchanged from place to place: it is a mediator.[68] As such, it is an 'actor', in the sense defined by 'actor-network theory', defined as 'not the source of an action but the moving target of a vast array of entities swarming towards it'.[69]

The nineteenth century produced an intensification of the condition known as modernity in which, according to one form of its self-understanding, human beings were both dislocated from the natural world of matter, and swallowed up in a 'materialism' of their own making. Matter was first abstracted from nature, then, as a result, nature – including human nature – was itself abstracted into matter, set intimidatingly and puzzlingly over against human life. In either case, it is believed that what is lost is some mutually defining and enriching interchange between matter and whatever exceeds it, in the manner of Emerson's imagination. Now, matter can only be figured in symbolism or representation. Bruno Latour has urged us to do without this myth of dissociation, paying attention instead to an 'associology', or a 'science of the living together' of humans and non-humans.[70] This will yield us a vision, not of tongue-wagging subjects set against mute, mysterious objects, nor a dialectical overcoming of their antagonism, but rather an attentiveness to the 'Middle Kingdom', in which quasi-objects and quasi-subjects reciprocally precipitate each other:

> Of quasi-objects, quasi-subjects, we shall simply say that they trace networks. They are real, quite real, and we humans have not made them. But they are collective because they attach us to one another, because they circulate in our hands and define our social bond by their very circulation.[71]

Carefully regulated by a series of joints, relays, switches and valves, gas itself became a mobile switch – between the global and the local, the economic and the physiological, the immediate and the mediate. It is a matter that ultimately enacted a translation or switching between matter and the immaterial. As such, it provides a kind of correlative to the imaginary plasma of the novel itself, which equally arises from a curiosity with the complexity of conjunctures, with the intimate traffic of proximity and distance. In this respect, the novel itself may be seen as one of the nineteenth century's most pervasive gas works.

five

Transported Shiver of Bodies: Weighing Ether

The ether is a thoroughly nineteenth-century notion, and the period during which it was toiled and tussled over by physicists fits very neatly into the frame provided by the century. The nineteenth-century career of the ether began as the century itself clocked on, with the experiments by Thomas Young in 1801 that demonstrated interference patterns produced by two beams of light, which suggested that light moved in the form of waves. It was not easy for Young's theory to make headway in Britain against the corpuscular theory of light that predominated there, but his work was developed and verified over the next two decades in France by Augustine Fresnel and François Arago. If, as Boyle and others had amply demonstrated, light travelled through a vacuum with no observable diminution of speed, then it seemed reasonable, and indeed necessary, to assume that there was some medium to carry the light. Young favoured a stationary ether, one completely unaffected by the motion of material bodies or the passage of waves through it: 'I am disposed to believe, that the luminiferous ether pervades the substance of all material bodies with little or no resistance, as freely perhaps as the wind passes through a grove of trees.'[1] By the middle of the century, the existence of this medium, universally diffused, unimaginably tenuous, but indispensably existent, was widely accepted, at least among British, which is in fact largely to say Scottish and Irish, physicists.

It is customary for historians of the ether preoccupation to date the beginning of its demise, at the other end of the century, from the so-called Michelson-Morley experiments of 1881 and 1887. Albert Michelson followed up a suggestion made by James Clerk Maxwell in his article 'Ether', in the ninth edition of the *Encyclopaedia Britannica*, that if there were an ether, then it could be expected to exert a drag on the earth as it moved through ether-filled space, similar to the drag exerted by a river on the hull of a boat. This ought to result in light travelling at slightly different speeds in different directions. Michelson devised and executed a superbly elegant experiment to test this hypothesis. There was no such variation: light travelled at precisely the same speed in all directions, indicating that there was no ether-wind to urge or retard its course depending on the direction it was trying to travel. Michelson's and Morley's measurements gave impetus to the appearance in 1905 of a paper modestly entitled 'On the Electrodynamic Properties of Moving Bodies' by one Albert Einstein. In suggesting that the space is not the simple and invariant container of matter, light, magnetism and gravity, but is itself subject to transformation as a result of them, the special theory of relativity is taken to have rendered the ether superfluous to requirements. Promptly, without protest, and with only the faintest expiring sigh, so the story goes, the ether 'ebbed quietly out of the universe early in the twentieth century'.[2]

As usual, this story of centenary infancy, heyday and demise is an artefact. William Thomson could still be heard declaring in 1884, in a popular lecture on the wave theory of light he gave in Philadelphia, that the 'luminiferous ether' was 'the only substance we are confident of in dynamics. One thing we are sure of, and that is the reality and substantiality of the luminiferous ether'.[3] Of course, it might be said that this was a kind of defiant, dying flare of the ether hypothesis, as theorists and experimenters tried desperately to respond to the Michelson-Morley challenge. In fact, though arguments and experiments were mounted to try to rescue the ether from the Michelson-Morley experiment – for example, the whirling ether machine built by Oliver Lodge to show that the earth carried portions of the ether along with it – the many writers who published on this topic during the 1890s show little attempt to respond to Michelson's and Morley's debunking of the ether, or even awareness of the experiments. So there is a sense after all in which, in cultural-epistemological terms at least, the ether does indeed display a very considerable drag coefficient. It is scarcely surprising that the ether should have survived into and recurred in the twentieth century too.

Indeed, there is a sense in which the theory of relativity may have naturalized and perpetuated some aspects of ether thinking, rather than dispelling it. The idea of the ether may lie behind and make possible some of the thinking about space that has become possible, not just in science, but in geography, architecture and cultural studies. The ether occupied the absolute space that Newton's theories required, space in other words conceived simply as a neutral frame or container within which the drama of cosmic force and matter could be played out. But conceiving ether as the matter of space, or space as a kind of matter, made that space torsive, mutable, productive. In the idea of ether, wrote Einstein, a quarter of a century after he was supposed to have evacuated it,

> . . . space gave up its passive role as a mere stage for [physical events] . . . The ether was invented, penetrating everything, filling the whole of space, and was admitted as a new kind of matter. Thus it was overlooked that by this procedure space itself had been brought to life.[4]

In one sense, the ether had to be dispensed with in order for space to be drawn into the relations of reciprocal transformation on which theories of relativity depended. In another sense, ether theories made that very transition possible.

Pondering

Most poetic and literary uses of the word 'ethereal' in the nineteenth century connote a traditional notion of unreality or fragile spirituality. But there was nothing in the least ethereal in this sense about the conceptions of the ether developed by nineteenth-century physicists. Assuming as they did that the ether was a mechanical as well as a philosophical necessity, nineteenth-century British physicists did everything they could to give mass and dimension to this enigmatic, and hitherto imaginary substance. The history of the ether is a history of the pondering of the imponderable. This pun is at the heart of ether-thinking, in which the question of whether the ether could indeed be regarded as having weight was central. Was it stationary, a notional frame or necessary nothing, a mere, minimal *that-through-which* optical and, later, electromagnetic waves might pass? Or was it a something, with its own extension, mass and ponderability? If so, did it act like a fluid? Or like a sponge? A jelly? A foam? Was it completely elastic, or, by contrast, almost entirely rigid?

Among the many theories of the ether that appeared during the 1890s, when the orthodoxy of the idea of the ether in physics produced many pseudo-scientific speculations entirely free of any mathematical or experimental support, was Alfred Senior Merry's bizarre view, expressed in his pamphlet *Interstellar Aether* of 1891, that the universe was made up of two elements, broadly identifiable as electricity and heat, which Merry called 'thermine' and 'electrine'. (There is no direct connection with the instrument called the theremin, which was named after its inventor Leon Theremin, though it is intriguing to note that it was originally known as the 'etherphone'.) Merry insisted that what was thought of as forces ought to be thought of as having material existence, and might even have mass, but pointed to an interesting difficulty arising from the idea that the ether was a kind of matrix within which the world of matter existed:

> [T]he general objection to this is that they are imponderable; but in the first place, as they are probably the *cause* of attraction of gravitation, there is no reason that they should be subject to the law which they themselves create; in the second place, they *may have* very great weight. But it has hitherto been impossible to prove this, as we have never been able to create a vacuum free from Electro-thermine; and therefore, as air cannot be weighed in air, or water in water, so also Electro-thermine cannot be weighed in Electro-thermine – for, as I admit, interstellar space is occupied by, and all matter permeated with it.[5]

The central problem that the hypothesis of an ether is designed to solve is how gravity works. If the ether perhaps transmits gravity in the same way as it transmits light, then what could it mean for it itself to have mass, and therefore to exert a gravitational force? The ether cannot originate or explain a force to which it is itself subject. The idea of the ether arises precisely out of an interference between weight and lightness.

The effort to try to conceive and characterize the ether as literal, actual and massy led to one of the most remarkable subsidiary episodes in the nineteenth-century history of the ether. In 1867 William Thomson witnessed a demonstration by his friend and fellow-physicist P. G. Tait of a machine for the production of smoke rings. Thomson was reminded of an essay by Hermann Helmholtz he had encountered through Tait on the stability of vortexes within water.[6] Smoke rings demonstrated the same kind of stability: although smoke rings are simply localized patterns of air flow, they behaved as though they had density and resistance: two smoke rings meeting each other would kiss and bounce away from each other like bubbles or billiard balls.

Thomson's intuition was that atoms might be regarded as this kind of whirling or torsion, though not within air or water, but within ether.[7] Thomson's theory of oscillating vortex-rings – or 'worbles', as James Clerk Maxwell called them, after Helmholtz's 'Wirbelbewegungen'[8] – was the most important of a number of theories that proposed that matter itself might be a kind of contusion or convolution in the ether. The ether thereby underwent a kind of ontological promotion. No longer merely that which lay between things, the inert and docile soup studded with beads of matter, the ether became primary, an ur-matter, or quasi-matter, out of which all things were made. One of the advantages of this view for more religiously orthodox scientists was that, though the stability of vortex rings in a frictionless medium would mean they would last eternally, as atoms seemed to, they could not be regarded, for that very reason, as able to arise through some accident or evolution in the state of the ether; they required an act of precedent creation. On the other hand, there were religiously orthodox scientists, like P. G. Tait, whose smoke rings had precipitated the ether-knot hypothesis in Thomson's mind, and whose influential book *The Unseen Universe*, written with Balfour Stewart and published anonymously in 1875, made the ether a central part of a religious argument. Stewart and

Tait were concerned at the requirement that a creator should have to have intervened to modify the state of a perfect fluid to create ether-vortices, and preferred to postulate originary wrinkles in the fluid to allow for their evolution.

For thirty years or more, long after the ether was supposed to have been rendered defunct by the Michelson-Morley experiments, British physicists attempted to develop models and equations that would substantiate this view that matter is formed from episodic knots in nothingness.[9] Such views may have been particularly popular among crackpots and cranks, but they also prepared the way for a view of matter that dispensed with the absolutes of Newtonian mechanics. Late in the 1920s John Wills Cloud was arguing in his *Castles in the Ether* and *The Ether and Growth*, that the ether consisted of tiny particles he called 'Ethrons', packed tightly together, like shot. Matter existed in the interstices between these ethrons. 'All matter is therefore in the ether, and the ether is in all matter, so that every material form is an image of the ether within it, in the sense of having the same outlines.'[10] Cloud went further to suggest that the ether was 'the sole source of all evinced mechanical energy and . . . provides the material for the genesis of matter'.[11]

Some have seen the ether as marking the final stage of classical physics, a last writhe of the naïve demand that the universe must be shown to be *made out of something*, some ultimate and universal substance, rather than arising immanently out of relations and reciprocities. This view could be used to assist a mechano-materialist theory like Tyndall's peculiarly rhapsodic evocation of the evolution of life, form and intelligence from disorganized, nebulous matter:

> the more ignoble forms of animalcular or animal life, not alone the nobler forms of the horse and lion, not alone the exquisite and wonderful mechanism of the human body, but that human mind itself – emotion, intellect, will, and all their phenomena – were once latent in a fiery cloud.[12]

The ether may have appeared in this light as a pulsing residue of this primal nebula, performing a function similar to that of the cosmic microwave background, at which contemporary cosmologists peer in the search for clues as to the originary constitution of the pre-explosive universe. For more traditional religious thinkers, such as William Whewell, the ubiquity and complexity of the ether was the sign of a transcendent divine purpose. Whewell's 1833 Bridgewater Treatise argued that the ether

> must not be merely like a fluid poured into the vacant spaces and interstices of the material world, and exercising no action on objects: it must affect the physical, chemical and vital powers of what it touches. It must be a great and active agent in the work of the universe, as well as an active reporter of what is done by other agents.[13]

The mere fact that there should be in the universe 'a machine as complex and artificial, as skilfully and admirably constructed' as the ether, Whewell continued, 'is well calculated to extend our views of the structure of the universe, and of the resources, if we may so speak, of the Power by which it is arranged'.[14] For Whewell, '[t]here is nothing in all this like any material necessity, compelling the world to be as it is and not otherwise'.[15] But the marvellous complexity of the ether is precisely what underpins John Tyndall's materialist argument, which suggests that, when nature itself displays this kind of complexity, there is no need to posit a creator. So, although it is true that, in Peter J. Bowler's words, '[t]he ether became a vehicle by which the universe could once more be seen as a unified whole with a purposeful structure', this very same attitude could also support a materialist view.[16] The materialist and the providentialist in fact seem to employ the same argument, about the extraordinary complexity of the effects flowing from what seems to be a primary and ubiquitous substance in the universe.

The ether could certainly be regarded as dissolving all certainties about matter and life in the universe. Oliver Lodge, who was to remain one of the staunchest defenders of the ether well into the 1920s, making it the central principle of his belief that physics could verify the truth of survival beyond death and communication with the dead, concluded in his Romanes lecture in Oxford in 1903 that 'the whole of existing matter appears liable to processes of change, and in that sense to be a transient phenomenon'.[17] Joseph Conrad was among those who seemed most shaken by the dematerializing power of the new physics. He wrote in a letter of 29 September 1898 to Edward Garnett:

> The secret of the universe is the existence of horizontal waves whose varied vibrations are at the bottom of all states of consciousness. [Matter is] is only that thing of inconceivable tenuity through which the various vibrations of waves (electricity, heat, sound, light, etc.) are propagated, thus giving birth to our sensations – then emotion – then thought.[18]

John Davidson, who probably read of the ether in Ernst Haeckel's popular book *The Riddle of the Universe*, similarly referred to it as 'omnisolvent ether'.[19]

From the late 1880s onwards, occultists and supernaturalists seized upon the idea of the ether, thereby becoming themselves a medium of transmission from specialized physics to popular understandings and, even more slowly, to modernist artists, the most advanced of whom tended to catch up with scientific ideas about two decades after the readers of popular science they so despised. As G. N. Cantor has shown, these occultists and supernaturalists found in the ether a materialist confirmation for the traditional assurances of religious and mystical thought that the universe was entire, harmonious and self-consistent.[20] Conrad concluded his reflections on waves with the statement that 'there is no space, time, matter, mind as vulgarly understood, there is only the eternal something that waves and an eternal force that causes the waves'.[21] In his presidential address to the British Association for the Advancement of Science in 1904, the Conservative politician Arthur Balfour freely conceded the principle that 'gross matter, the matter of everyday experience, [is] the mere appearance of which electricity is the physical basis', meaning that 'the beliefs of all mankind about the material surroundings in which it dwells are not only imperfect but fundamentally wrong'.[22] In the new physics, he quotably declared, 'matter is not merely explained, but is explained away'.[23] And yet, he argued, 'without the ether an electric theory of matter is impossible'.[24] Balfour thus joined many others in seizing on the ether as the underlying principle of continuity in the universe, the guarantee that, amid all this incessant tremulous dissolution, there was a principle of continuity, a bottom line, a *ne plus ultra*:

> Two centuries ago electricity seemed but a scientific toy. It is now thought by many to constitute the reality of which matter is but the sensible expression. It is but a century ago that the title of an ether to a place among the constituents of the universe was authentically established. It seems possible now that it may be the stuff out of which that universe is wholly built.[25]

In fact, one might see in the convolutions of thought and value about the ether in the period of early modernism an anticipation of modernism's own somersaulting attempt to found a system of value in immaculate mid-air, in the absolute absence of all absolutes.

Shiver of Bodies

We are accustomed to the idea that the nineteenth century was very concerned with the nature and status of matter. But the ether had another kind of dimension in this period. On the one hand, it was a simple convenience, called into being by the logical necessity for there to be some kind of medium through which light and other radiating impulses might travel. Seen in this way, the ether need have no qualities in itself, other than that of being able to transmit impulses and undulations. On the other hand, it is clear that these impulses do not simply pass through the ether like light through a window pane, for, whether the waves it transmits are longitudinal or transverse, the transmission of the wave is effected by local movements within the ether itself, just like the movements of individual particles of water that produce waves and ripples in the sea. This point of view sees the ether as actively present and involved in the transmission of the wave, rather than merely its vehicle or occasion. Physicists for the most part concerned themselves with the physical properties – optical, mechanical and, later, electromagnetic – of the alleged ether. But the extension of the notion of the ether into areas of medicine, the occult and popular science meant that it began to function more and more as a mode of sensitivity or susceptibility, as a kind of entity, or quasi-vital substance, rather than a form of matter – space, in Einstein's phrase, brought to life, in a line of thinking that reactivated the Stoic *pneuma*, the active fiery principle that pervades and animates the cosmos.

Many of the evocations of the ether stress its quasi-animate or animating nature. In such evocations, the ether is a kind of prototype for the sensitive flesh that it will slowly bring about. John Tyndall stressed the intimate and almost immediate bodily connection between the distant stars and the observer's eye:

> This all-pervading substance takes up their [the stars'] molecular tremors, and conveys them with inconceivable rapidity to our organs of vision. It is the transported shiver of bodies countless millions of miles distant, which translates itself in human consciousness into the splendour of the firmament at night. (*Fragments of Science*, 4)

Tyndall's 'transported shiver of bodies' might be thought of not just as putting bodies in contact with one another, but also as bringing bodies into being in the first place, as electricity induces a magnetic field and vice versa.

These ideas resonate strongly in the work of Walter Pater, whose 'Conclusion' to *The Renaissance* (1868) is full, not just of images of inconstancy, but also of regular pulsation: it evokes 'intervals', 'recoil'; the waste and repairing of the brain under every ray of light and sound', 'that strange, perpetual, weaving and unweaving of ourselves', under the influence of 'impressions' that do not merely cascade upon the sensibility of the subject, but oscillate, with inconceivable rapidity.[26] Pater's pulsations are also anticipated by Emerson, who insisted in his essay 'Experience':

> Man lives by pulses; our organic movements are such; and the chemical and ethereal agents are undulatory and alternate; and

> the mind goes antagonizing on, and never prospers but by fits. We thrive by casualties.[27]

Quality comes down to quantity, intensity to an effect of number, life consisting in the end merely of 'a counted number of pulses', an accountancy of intensity.[28] The nineteenth century began to make substantial and quantifiable an earlier theory of the unity of sensations, the distinctions between the senses being simply in the range of frequencies and periodicities to which they were sensitive.

This way of thinking, reviewed by Roger French in his 'Ether and Physiology' (1981), locates the ether, not just in interstellar space, but also in the human body, conceived of, not as fixed form or clear outline, but as a tremulous cloud of sensations and reflexes.[29] The idea of explaining the senses and sensitivities of the body in terms of etheric fluids has its origins in the doctrine of the animal spirits inherited from humoral psychology, and has a more specific source in the etheric fluid posited by David Hartley in his *Observations on Man* of 1749. Hartley suggested that all sensation has its basis and physical form in vibrations, imparted to and by a subtle fluid running through the nerves. He notes, indeed, that his notion of the ether derives from Newton's speculations in his *Principia*.[30] Hartley's ideas seem to have spilled out of the narrow context of his associationist psychology, and out of the nerves into the spaces between human bodies.

The propagation of the two waves of mesmerism-theory, during the 1780s and early decades of the nineteenth century, and then again between 1840 and 1880, depended on the belief in an ether-like magnetic effluvium, which pervades the universe, but is also concentrated in and emanates from living beings. Rather than an ether in which we are immanent or by which we are assailed, the magnetic effluvium was a kind of voluntary ether, or ether-on-elastic, which could be extruded, stored, concentrated and manipulated. As we saw in chapter Two, even after the apparently authoritative demonstration by a commission appointed by Louis XVI in 1784 that there was no such fluid, there was a strong tendency to identify it with the elastic ether, and the return of ether theories in the nineteenth century contributed significantly to the revival of mesmerism. J. S. Grimes, a theorist of phreno-mesmerism, the curious blend in which mesmerism was revived in the middle of the nineteenth century, called the magnetic fluid 'etherium', and the study of its effects 'etherology'.[31]

In 1874 Benjamin Ward Richardson, a prolific writer on a huge range of sanitary and temperance questions, attempted a soberer revival of the idea of the nervous ether. Within all living creatures, he proposed, there was an agency that mediated between the matter of which bodies were composed and the forces to which they were subject. This agency was the conductor of the vibrations of heat, light, sound, electromagnetic impulses and mechanical friction. Richardson envisages this substance as 'a finely diffused form of matter, a vapor filling every part, and even stored in some parts; a matter constantly renewed by the vital chemistry; a matter as easily disposed of as the breath, after it has served its purpose'.[32] It functions as

> an intercommunicating bond which connects us with the outer world; which is apart from the grosser visible substances we call flesh, bone, brain, blood . . . which receives every vibration or motion from without, and lets the same vibrate into us, to be fixed or reflected back; and which conveys the impulse when we will an act and perform it.[33]

High Orgasm

Most histories of the concept of the ether have been conducted in terms of the physics that first deemed it necessary, and then dispensed with it. But we have begun to see that there are other contexts and usages that could not help but drift into thinking about the ether, considered as a broader cultural practice of thought. William Thomson is reported as having said in 1896 to George Fitzgerald: 'I have not had a moment's peace in respect to electromagnetic theory since November 28, 1846. All this time I have been liable to fits of ether dipsomania, kept away at intervals only by rigorous abstention from thought on the subject.'[34] Thomson's little joke is made possible by the practice of what was called 'ether-drinking', which was still prevalent among 1890s decadents and bohemians such as Jean Lorrain, whose *Sensations et*

Souvenirs of 1895 has been translated as *Nightmares of an Ether Drinker* (2002).[35] It suggests a connection between the inebriation of the ether idea and the more literal kinds of intoxication by the substance that, not entirely by coincidence, shares its name.

Benjamin Ward Richardson is at pains to point out that, in referring to his mooted medium of nervous action and response as an 'ether', he means no reference to the chemical substance of that name, but uses the term rather 'as the astronomer uses it when he speaks of the ether of space, by which he means a subtle but material medium, the chemical composition of which he has not yet discovered'.[36] But in fact, Richardson would go on shortly afterwards to take a close interest in the effects of ether-intoxication, conducting an investigation at first hand of the epidemic of ether-drinking around Draperstown in Northern Ireland and subjecting himself experimentally to its effects. That there is a more than verbal coincidence between the ether of space and chemical ether is suggested by his view that, since the nervous ether is best considered as a kind of gas or vapour, it might be subject to contamination:

> Through the nervous ether, itself a gas or vapor, other gases or vapors may readily and quickly diffuse . . . Thus those vapors which, being diffused into the body, produce benumbing influence – as the vapors of alcohol, chloroform, bichloride of methylene, ethyllic ether, and the like – produce their benumbing effects because they are not capable of taking the place of the natural ether into which they diffuse; they interfere, that is to say, with the physical conduction of impressions through what should be the pure atmosphere between the outer and the inner world. A dense cloud in the outer atmosphere shall shut out any view of the sun; a cloud in the inner atmosphere of my optic tract shall produce precisely the same obscurity.[37]

The substance known as ether, more precisely diethyl ether, which is produced from a combination of distilled alcohol and sulphuric acid, was first distilled and described by the German scientist Valerius Cordus in 1540, while Paracelsus described its hypnotic effects at around the same time. It was known as 'sweet vitriol' until 1730, when W. G. Frobenius gave it the name 'spiritus aetherius', which yielded in turn to its more common name.[38] But the difficulty of producing it reliably meant that it did not come into common use until the mid-eighteenth century. Its most obvious property was its extreme volatility. A quack pamphlet of 1761 by Matthew Turner extols it as

> the most light, most volatile, and most inflammable, of all known Liquids: It swims upon the highest rectified Spirit of Wine as Oil does upon Water, and flies away so quickly as hardly to wet a Hand it is dropped upon; from which Properties it probably obtained it's Name. It is so readily inflammable, as to take Fire at the approach of a Candle, before the Flame reaches it. Any Electrified Body will also produce the same Effect.[39]

Turner also claimed it as a powerful solvent, with particular uses in dissolving gold, which was frequently drunk for medicinal purposes:

> It has a greater Affinity with Gold than *Aqua Regia* has . . . thus a true and safe *Aurum potabile* is readily prepared for those who want such a medicine. The Union of these two Substances is very remarkable, one being the heaviest solid Body we know, and the other the lightest Liquid.[40]

Recreational ether-drinking (so called, though in fact ether-sniffing or inhalation was an equally popular form of intake) began on a serious scale only after its anaesthetic properties were discovered in 1846 by Thomas Morton, who marketed it as 'letheon'. Germany was swept by 'etheromania', and there were other epidemics in Michigan and Lithuania. When Benjamin Ward Richardson tried it on himself in the 1870s, he experienced sensations of attenuation and lightening, as though the substance was capable of imparting its own diffusive qualities to those who took it in:

> [P]eriods of time were extended immeasurably . . . the small room in which I sat was extended into a space which could not

> be measured . . . the ticking of the clock was like a musical clang from a cymbal with an echo.[41]

He also recorded an odd sensation that suggests an involuntary invocation of his own theory of the mediating nervous ether: 'all things touched felt as if some interposing, gentle current moved between them and the fingers'.[42] The ether displays the same ambivalence as the astronomical ether, for it both enlarges sensibility, and yet acts as a mediator or cushion for it. Awareness is both 'spaced out', and brought into intimate contact with everything.

The mid-century indulgence in ether recapitulated the craze for the inhalation of nitrous oxide that was a feature of the turn of the nineteenth century. Nitrous oxide would be closely linked with ether during the 1840s, when their shared anaesthetic properties came to prominence. Though the nineteenth century experienced many new forms of gas and vapour (of which, as we saw in the last chapter, gas lighting was among the most important and pervasive) as well as important new ideas about gases, most notably James Clerk Maxwell's statistical explanations of their behaviour, it also inherited a complex and widely diffused idea of what might be called a 'pneumatic sublime' from Romanticism, and the great scientists of the Romantic period, Joseph Priestley and Humphry Davy, both of whom made their reputations in the study of gases. This sense of the authority and fascination with the vaporous in this period would lead T. E. Hulme to characterize Romanticism as 'always flying, flying up into the eternal gases'.[43] Nitrous oxide was the somewhat grotesque literalisation of the principle of airiness that is to be found throughout Romanticism – the inspiration of wind, the power of soaring ascent, the force of diffusion and the diffusion of force. Not that this is the first time that gas has been linked to prophecy – after all, the pythian priestess at Delphi had been reputed since the second century to derive her powers of prophecy from a vapour ascending from a crack in the earth.

Late nineteenth-century supernaturalism, with its apports and levitations, usually of conspicuously heavier-than-air subjects, such as the extensive Mrs Grundy or the spacious Madame Blavatsky, solidified and domesicated this fantasy of the pneumatic sublime, and

'Laughing Gas', line drawing after George Cruikshank by John Scoffern, *Chemistry No Mystery; or, A Lecturer's Bequest* (London, 1839).

popular entertainment was quick to tune in as well. Robert Houdin, the performer from whom the anti-spiritualist Houdini would take his name, explained the Indian Conjuror's illusion in which his son, Auguste Adolphe, appeared to sit on air, as an effect of the imbibing of ether. 'When this liquid is at its highest degree of concentration', he solemnly mock-explained, 'if a living being breathes it, the body of the patient becomes in a few moments as light as a balloon'.[44]

Intoxicants are part of the history of the material imagination, or ongoing cultural invention of matter, and invention of itself through it. The idea of the ether cannot be thought of without including the dream of the ethereal that is focused on intoxicating gases, even though this has been assumed to be a distraction or irrelevance to most historians of ideas. The thought of the paradoxical substance called ether is also a materialisation of thought itself, which, insofar as it is necessary to render matter truly intelligible, is never merely 'cultural'. Davy's discovery of the capacity of an aeriform substance to transform the texture of thought itself, along with his resulting intuition that the universe may be composed of ideas and impressions, is perhaps something more than a delusion, given that it anticipates the imbrication of matter and thought that has become a theme of quantum physics.

Suffiction

If the ether was far from down and out in 1905, it is also the case that the beginning of speculation about the ether is to be found long before the beginning of the nineteenth century, and this history exerts a significant pull on nineteenth-century thought. Even leaving aside the long history of speculation about the fifth element, or quintessence, of ether in Aristotle, and the Stoic doctrine of *pneuma* that derives from it, much the same reasoning that had made Aristotle and other plenarists of his kidney reject the possibility of voids in nature had led Newton to the supposition of some kind of universal medium. Newton's case is interesting. He was extremely reluctant to believe in action at a distance, and the whole system of his mechanics depended upon a universe governed regularly by the application of forces. But all the forces required a medium through which to be transmitted; even though the existence of such a medium, however tenuous and almost-there, threatened to ruin all the equations (in the end, Newton realized, the drag exerted by the ether would bring all celestial motions grinding to a halt). The ether question was a kind of irritation for Newton; he refused, as he famously wrote, to 'frame hypotheses' (though his phrase *hypotheses non fingo* has also been translated as 'I feign no hypotheses').[45] But Newton's refusal is itself feigned, since he did keep on framing hypotheses regarding the nature of the medium within which cosmic forces operated. For Newton, the ether was always a kind of thought experiment: an experiment in thought, which was also an experiment on thought. In the conclusion to his *Aids to Reflection*, Coleridge intuited the uneasiness of both Newton and his followers when he wrote that the supposition of the ether was to be suspected 'not only as introducing, against his own Canons of Right Reasoning, an *Ens imaginarium* into physical Science, a Suffiction in the place of a legitimate Supposition; but because the Substance (assuming it to exist) must itself form part of the Problem, it was meant to solve'.[46] The ether is a kind of test of the powers of visualization and the practice of 'seeing feelingly'. What may impress us about nineteenth-century writings about the ether is how much they depend upon a kind of dynamic imagination, focused not so much on how things appear as forms, as on how they felt and worked, as actions and stresses. One of the most important differences between the ether casually evoked by poets during the period and that ether that formed the subject of science was that the physical ether was subject to and itself transmitted force, stress and strain. Herbert Spencer, who sought to deduce the evolution of the universe out of fundamental laws of motion, emphasized in his *First Principles* that the ether was in part a projection of ideas about the experience of force and resistance. Both William Thomson and James Clerk Maxwell, leading figures in the development of nineteenth-century ether theory, were strongly motivated by the principle that, unless one can generate a mental or physical model of how a problem works, no amount of mathematical reasoning can suffice to explain it. ('What's the go o' that?' was apparently the favourite question of the inquisitive young Maxwell.[47]) And yet, the very literalism of ether theory, its very amenability to mind and muscle – one of the strongest adherents of the vortex-ether theory and subsequently the discoverer of the electron, J. J. Thomson, wrote of the importance of theories which can be 'handled by the mind'[48] – suggested that it might be a comfortable illusion, generated by beings laughably determined to make cosmic mechanics work like the familiar push-me-pull-you mechanics operating on earth.

This meant that thought about the ether was always liable to tip over into thought about the nature and status of thought itself. Because nobody was ever likely to be able to see the ether, or bottle it up conveniently for analysis, the ether could not but become a kind of allegory of the scientific imagination itself, as it attempted to stretch, contort, refine or volatilize itself and its own ideas of what matter was. Tensile metaphors are particularly prominent, for example, in the work of John Tyndall. 'The Constitution of Nature', the first lecture of his *Fragments of Science for Unscientific People*, evokes the 'incessant dissolution of limits' (*Fragments of Science*, 3). Gillian Beer represents this as a remark about the effects of radiation on objects in the world, but Tyndall is in fact here referring to the action of the mind going beyond the idea of limit, not of nature surpassing its own limits (unless, of course, one takes the mind of man to be one of the ways in which nature reaches beyond itself). For the conception of the ether is itself an effect of this extrapolation, or radiation of the mind:

> Men's minds, indeed, rose to a conception of the celestial and universal atmosphere through the study of the terrestrial and local one. From the phenomena of sound as displayed in the air, they ascended to the phenomena of light as displayed in the *aether*. (*Fragments of Science*, 3–4)

Tyndall reverts to this idea in his 1870 lecture 'On The Scientific Use of the Imagination', again in reference to the extrapolation from sound to light and deduction of the existence of the luminiferous ether: 'There is in the human intellect a power of expansion – I might almost call it a power of creation – which is brought into play by the simple brooding upon facts' (*Fragments of Science*, 133) How are we to explain the rapid velocity of light, compared with the slowness of sound? Again, says Tyndall, it is in a mimicking of the process of radiation itself, '[b]y boldly diffusing in space a medium of the requisite tenuity and elasticity' (*Fragments of Science*, 134). The idea of the ether is therefore the proof of the imagination's power of going from sensual to supersensual fact; the idea of the ether is itself a phenomenon of radiation: 'In forming it that composite and creative unity in which reason and imagination are together blent, has, we believe, led us into a world not less real than that of the senses, and of which the world of sense itself is the suggestion and justification' (*Fragments of Science*, 134)

Modern physics requires the supplementation of observation by a kind of dynamic imagination that Tyndall always represents as expansive:

> Iron is strong; still, water in crystallising will shiver an iron envelope, and the more unyielding the metal is, the worse for its safety. There are men amongst us who would encompass philosophic speculation by a rigid envelope, hoping thereby to restrain it, but in reality giving it explosive force. (*Fragments of Science*, 37)

Later on he says that '[t]o it [the speculative mind] a vast possibility is in itself a dynamic power' (*Fragments of Science*, 158). The idea of radiation is itself radiant. The work of thought redoubles and extends the expansive work of matter. The link between the ether and the mind is also suggested by the fact that, in his 1865 essay 'Radiation', Tyndall calls the conception of the ether 'the most important physical conception that the mind of man has yet achieved' (*Fragments of Science*, 176).

Elsewhere, Tyndall can find other natural forms embodied in the imagination: the imagination can not only radiate or propagate, it can also focus, or be concentrated, like an atom. What do you reach at the end of your orderly speculations about the source of aether waves?

> The scientific imagination, which is here authoritative, demands as the origin and cause of a series of aether waves a particle of vibrating matter quite as definite, though it may be excessively minute, as that which gives origin to a musical sound. Such a particle we name an atom or a molecule. I think the seeking intellect when focussed so as to give definition without penumbral haze, is sure to realise this image at the last. (*Fragments of Science*, 136)

Tyndall could even draw the physical circumstances of his lecture, which was originally given as an after-dinner speech at the British Association in Liverpool into this reflexive circle. Evoking his nervousness at addressing so august a company he says: 'My condition might well resemble that of the aether, which is scientifically defined as an assemblage of vibrations' (*Fragments of Science*, 140).

Herbert Spencer also maintained a strong conception of the ether, taking the existence and effect of ethereal undulations for granted throughout his work. As the bearer of vibrations – or the ur-form of the vibration, since it is so very hard to know in what the ether consists apart from the vibrations it transmits – the ether is a kind of symbol of what, in a letter to his father of March 1858, he called 'the *universality of rhythm*; which is a *necessary* consequence of the antagonism of opposing forces. This holds equally in the undulations of the etherial medium, and the actions and reactions of social life'.[49] In 'The Filiation of Ideas', an 1899 account of his own intellectual evolution that appeared as an appendix to David Duncan's biography of him, Spencer describes how he came in 1857 to his sudden understanding of the essential similarity within propagation of all such forms of regularly-alternating movement:

> During a walk one fine Sunday morning (or perhaps it may have been New Year's Day) in the Christmas of 1857–8 I happened to stand by the side of a pool along which a gentle breeze was bringing small waves to the shore at my feet. While watching these undulations I was led to think of other undulations – other rhythms; and probably, as my manner was, remembered extreme cases – the undulations of the ether, and the rises and falls in the prices of money, shares and commodities. In the course of the walk arose the inquiry – Is not the rhythm of motion universal? And the answer soon reached was – Yes . . . As, during the preceding year, I had been showing how throughout all orders of phenomena, from nebular genesis to the genesis of language, science, art, there ever goes on a change of the simple into the complex, the uniform into the multiform, there naturally arose the thought – these various universal truths are manifestly aspects of one universal transformation. Surely, then,

> the proper course is thus to exhibit them – to treat astronomy, geology, biology, psychology, sociology and social products, in successive order from the evolution point of view.[50]

The ether is not just the mainspring, or hidden motive principle for all these parallel forms of oscillation: it is a model for the very form of inductive analogy that allows Spencer to move between different areas of thought.

The ether could provide a model not just for the mind's powers of radiation, but also for its oscillatory nature. Spencer puts into the mouth of a 'materialist of the cruder sort' in his *Principles of Psychology* of 1880 some subtle reflections on the oscillation between the ponderable and the imponderable.

> Far greater community than this has been disclosed between the ponderable and the imponderable: the activities of either are increasingly modified by the actions of the other. Each complex molecule of matter oscillating as a whole – nay, each separate member of it independently oscillating – causes responsive movements in adjacent ethereal molecules, and these in remoter ones without limit while, conversely, each ethereal wave reaching a composite molecule, changes more or less its rhythmical motions, as well as the rhythmical motions of its component clusters and those of their separate members.[51]

These reflections themselves demonstrate the principle of oscillation, between the either/or of ether and matter, and between this either/or and another, the either/or of matter and thought about matter. The object of theory is an allegory of it.

These rhetorical effects in the writing of science shadow the anthropomorphic vitalism that is a feature of occultist and spiritualist apprehensions of the ether. The notion of the ether comes into being from the very dynamic activity of the mind that it seemed to quicken, and with which, for some, it was identical. The ether was an act of mind; to think of it is to do nature's own work of thought.

Aetherial Commotion

The most important development in ideas of the ether in the second half of the nineteenth century was its implication in the transmission of electromagnetic impulses. Following James Clerk Maxwell's predictions, Heinrich Hertz produced the first radio wave in 1887. Marconi's study of Hertzian wave theory led him to his first broadcast of a radio signal in 1895. To conceptions of the luminiferous ether, the mechanical ether, the nervous ether and the electromagnetic ether was added what we might call the 'informational ether' of the communications revolution that began in the late nineteenth century.

As originally conceived, the ether was a sort of embodied nothing, a form of matter as close to vacuity as it was possible to be. As the ether multiplied its functions, it appeared to become more and more crowded. Tyndall had already emphasised how full space might seem when one considered the variety of undulatory motions it could transmit without interference:

> In the spaces of the universe both classes of undulations [heat and light] incessantly commingle. Here the waves from uncounted centres cross, coincide, oppose, and pass through each other, without confusion or ultimate extinction. The waves from the zenith do not jostle out of existence those from the horizon, and every star is seen across the entanglement of wave motions produced by all the other stars. It is the ceaseless thrill which those distant orbs collectively create in the aether, which constitutes what we call the *temperature of space*. As the air of a room accommodates itself to the requirements of an orchestra, transmitting each vibration of every pipe and string, so does the inter-stellar aether accommodate itself to the requirements of light and heat. Its waves mingle in space without disorder, each being endowed with an individuality as indestructible as if it alone had disturbed the universal repose. (*Fragments of Science*, 178)

Spencer too evoked the amazing complexity of a substance that was able both to transmit everything and yet also to maintain distinctions and sympathies. In such a universe, distance and proximity no longer have a simple spatial meaning:

> The discovery that matter, seemingly so simple, is in its ultimate structure so amazingly involved, the discovery that, while it appears to be inert, it is the seat of activities immense in quantity and complication; and the discovery that its molecules, pulsating with almost infinite rapidity, propagate their pulses into the all-surrounding ether which carries them through inconceivable distances in infinitesimal times; serve to introduce us to the yet more marvellous discovery that molecules of each kind are specially affected by molecules of the same kind existing in the farthest regions of space. Units of sodium on which sunlight falls, beat in unison with their kindred units more than ninety millions of miles off, by which the yellow rays of the sun are produced. Nay, even this is a totally inadequate illustration of the sympathy displayed by the matter composing the visible Universe. The elements of our Earth are thus connected by bonds of interdependent activity with the elements of stars so remote that the diameter of the Earth's orbit scarcely serves as a unit of measure to express their distances.[52]

Far from being inert, the ether 'is never still', wrote Tyndall. 'To the conception of space being filled, we must therefore add the conception of its being in a state of incessant tremor' (*Fragments of Science*, 8). His evocation of the state of 'aethereal commotion' (*Fragments of Science*, 8) in the midst of which we live on the surface of earth anticipates the sense that began to grow from the 1890s onwards that human beings were diffusing more and more of themselves into the ether.

Once again, this ethereal commotion seems recursively self-designating. The cultural space designated as 'the ether' was similarly traversed by many different undulations and wavelengths. These traversals are semi-comically treated in Kipling's 1902 story 'Wireless'. In this, a young, consumptive man is given a drink of chloric ether in a chemist's shop in which an experiment in radio transmission is to take place. Either the chemical ether, or the waves being transmitted through the electromagnetic ether, combine to induce him to entranced automatic writing, in which he seems to channel the words of the young Keats.[53] The point of the story is to demonstrate the possibility of interference between the different registers of the ether: chemical, electromagnetic, poetic and spiritual.

Perhaps the ether is always this kind of old–new compounding. The physico-poetico-magico-psychopharmacopia of ether-thinking in the nineteenth century and beyond is not a mere epoch of mind, but a saturated space in which the mentality of matter and the materiality of mind set up unavoidable and fertile interference patterns, not least, as we shall see in the next section, in the new visibility and audibility of the new idea of interference itself.

PART TWO

Atmospherics

six

Haze

Culture and the Weather

In *The Satanic Verses*, Salman Rushdie has Gibreel Farishta reflect on the ways in which 'the moral fuzziness of the English was meteorologically induced', and enumerates the benefits that might arise from a metamorphosis of London into a tropical city:

> . . . increased moral definition, institution of a national siesta, development of vivid and expansive patterns of behaviour among the populace, higher-quality popular music, new birds in trees (macaws, peacocks, cockatoos), new trees under the birds (coco-palms, tamarind, banyans with hanging beards). Improved street-life, outrageously coloured flowers (magenta, vermilion, neon-green), spider-monkeys in the oaks . . . better cricketers; higher emphasis on ball-control among professional footballers, the traditional and soulless commitment to 'high workrate' having been rendered obsolete by the heat. Religious fervour, political ferment, renewal of interest in the intelligentsia . . . Spicier food; the use of water as well as paper in English toilets; the joy of running fully dressed through the first rains of the monsoon.[1]

In general, we find this kind of parallel between culture and weather either pleasantly absurd, or true only in so general or folkish a sense as to lack any real interest or utility. It is not that there are no intelligible parallels between conditions of weather and conditions of the soul – the many climatic terms we have to designate the vicissitudes of personal and national character see to that. It is that the parallels are too well established. Culture and the weather seem to run on parallel tracks, each a metaphor for the other, that only rarely and gratuitously intersect as effects rather than images of each other. Culture and weather are different orders, connected only by mediations so vast and complex that no non-trivial determinations can really be established. The weather has temporality – we say it is all tempestuous temporality – but no history. The time of the weather is a time without retention. It is pure fluctuation, without pattern, memory or history, movement without duration or direction or progression. Human affairs are historical in the sense that they are bound in protention and retention: the past is actively involved in the present, and the future is an active production of the present. One may chronicle the weather to be sure; but weather has no history in this sense.

One of the rare attempts to offer something like a meteorological history is Virginia Woolf's *Orlando*. At the end of chapter Four, Woolf gives us an historical transition rendered in meteorological terms. Orlando, now a female inhabitant of the eighteenth century, looks back on the 'huddle and conglomeration' of the previous century:

> A white haze lay over the town, for it was a frosty night in midwinter and a magnificent vista lay all round her. She could see St Paul's, the Tower, Westminster Abbey, with all the spires and domes of the city churches, the smooth bulk of its banks, the opulent and ample curves of its halls and meeting-places. On the north rose the smooth, shorn heights of Hampstead, and in the west the streets and squares of Mayfair shone out in one clear radiance. Upon this serene and orderly prospect the stars looked down, glittering, positive, hard, from a cloudless sky. In the extreme clearness of the atmosphere the line of every roof, the cowl of every chimney, was perceptible; even the cobbles in the streets showed distinct one from another, and Orlando could not help comparing this orderly scene with the irregular and huddled purlieus which had been the city of London in the reign of Queen Elizabeth.[2]

Her retrospect anticipates a cinematic move that announces the arrival of a new century. On the stroke of midnight, a small cloud appears behind St Paul's and begins to spread across the sky:

> [T]he cloud spread north. Height upon height above the city was engulfed by it. Only Mayfair, with all its lights shining, burnt more brilliantly than ever by contrast. With the eighth stroke, some hurrying tatters of cloud sprawled over Piccadilly. They seemed to mass themselves and to advance with extraordinary rapidity towards the west end. As the ninth, tenth, and eleventh strokes struck, a huge blackness sprawled over the whole of London. With the twelfth stroke of midnight, the darkness was complete. A turbulent welter of cloud covered the city.
>
> All was darkness; all was doubt; all was confusion. The Eighteenth century was over; the Nineteenth century had begun.[3]

Here, it is the very legibility of cultural-historical and meteorological parallels that make them instances of pathetic fallacy. And yet, even to put it this way suggests that we may be entering a different order, one in which, as Michel Serres has put it in *The Natural Contract*, 'global history enters culture; global culture enters history: this is something utterly new in philosophy'.[4] So is the fact of nature entering history a natural fact, or an historical fact?

I want to sift through some instances of the engagement among modernist artists with a particular meteorological phenomenon. The naming of weather phenomena is always approximate, but I have given my subject the term 'haze', though I mean this designation to spread, hazily enough, over other phenomena like mist, fog and smog. As of indistinction itself, it is not possible to distinguish absolutely the features and functions of haze – to draw the line between mist, haze, fog. I mean nevertheless to try to make out the terms of a general affinity between modernism and the nebular. I will try to show that modernist haze was a phenomenon not just of ambivalence, but, more exactly, of interference, an accidental mixing of registers and channels. It is a kind of visual noise, which implicates the conditions of perception and registration in its nature. While modernist painters and writers sought to capture the effects of haze, to make visible the forms and effects of indiscernibility and compromised vision, it was never clear what it would mean to get a fix on that shifting dimnness, to get the unfocused in perspective. In one sense, haze was the residue of the past, which threatened to dissolve all distinctions, thwarting the subject's attempts to achieve clarity and distinctness. But the very indistinctness of haze was also, for a significant number of modernist artists, an objective, a vocation and a provocation.

Traditions of the Air

Modernism inherits two traditions or sets of associations with regard to haze. First of all, there is Romantic haze, which does not belong entirely or securely to the period we know as Romanticism. This is the haze of glamour, or diffused radiance. It is governed by what, in my *Book of Skin*, I characterized as the logic of the aura.[5] The logic of the aura is that what spills out from the body of persons or entities of particular sanctity or spiritual power is also retained by it; the aura forms a second skin, or series of such skins, which remains held in by, and obedient to, the contours of the first. The aura is an emanation that, like the logos, goes forth from and yet also remains, and remains in, itself; as the Kabbalistic Zohar says, it breaks out and yet does not depart. It also sets that origin apart, keeps it at a reverent distance. The idea of a shining mist is a magical compromise between two principles – the scattering or diffraction of light, and the gathering or lingering of that light in a visible form. A mist both diffuses and detains radiance.

In Matthew Arnold's 'The Youth of Nature', haze is imbued with this kind of richness, which keeps the diffused spirit of Wordsworth intact:

> The mountains stand at its head
> Clear in the pure June-night,
> But the valleys are flooded with haze.
> Rydal and Fairfield are there;
> In the shadow Wordsworth lies dead.
> So it is, so it will be for aye.[6]

There is another tradition of the vaporous, which accounts for traditions of will o' the wisps, and other such atmospheric mirages. According to this tradition, perception is endangered by the exhalations from the ground, just as bodily health is. Dreams and delusions are contaminations of the pure, crystalline transparency of eye and mind. Britain was notorious for the haziness of its atmosphere, though Alexander Pope thought Ireland even worse, and the prospect of living there prompted in him an unusual sense of the word 'atmospherical', as meaning subject to atmospheric influences: 'If I liv'd in Ireland, I fear the wet climate wou'd indanger more than my life, my humour, and health, I am so Atmospherical a creature.'[7]

This tradition of hostility towards the corrupt and corrupting nether air re-emerges in the concerns about fog, smog and other airborne emanations in urban environments. There was little mellowness in the season of mists that affected urban environments in the nineteenth century. Nineteenth-century depictions of fog, the most extended and uncompromising of which is, of course, Dickens's at the beginning of *Bleak House*, inherit the vaporous sensibility of the medieval and late modern world, for whom mists and fogs are held to be the unhealthy halitosis of the ground, constituting a funerary air, full of infection, as opposed to the ethereal lucidity of the upper air. Where the radiant haze broke out from objects and persons of unusual vitality, and marked off their singularity, Dickens's fog does not allow for distinctions, but creates a kind of universal association in non-identity:

> Fog everywhere. Fog up the river, where it flows among green aits and meadows; fog down the river, where it rolls defiled among the tiers of shipping and the waterside pollutions of a great (and dirty) city. Fog on the Essex marshes, fog on the Kentish heights. Fog creeping into the cabooses of collier-brigs; fog lying out on the yards and hovering in the rigging of great ships; fog drooping on the gunwales of barges and small boats. Fog in the eyes and throats of ancient Greenwich pensioners, wheezing by the firesides of their wards; fog in the stem and bowl of the afternoon pipe of the wrathful skipper, down in his close cabin; fog cruelly pinching

> the toes and fingers of his shivering little 'prentice boy on deck. Chance people on the bridges peeping over the parapets into a nether sky of fog, with fog all round them, as if they were up in a balloon and hanging in the misty clouds.[8]

Fog is the undoing of place and spatial differentiation. The fog creeps in and out of every crevice, and turns the earthly into the airborne.

Vaguer in the Air, More Soluble

Many modernist artists defined their impulses and ambitions in terms of a desire for clear definition and a hostility towards the gaseous. To make it new was to burn off the dim mists of faerie. Ezra Pound memorably counselled an art of crystalline hard edges and outlines, advising in 1913: 'Don't use such an expression as "dim lands of peace". It dulls the image. It mixes an abstraction with the concrete. It comes from the writer's not realizing that the natural object is always the *adequate* symbol.'[9] If the cloud that spreads over London in chapter four of *Orlando* suggests that the nineteenth century could be seen as one long, clammily unremitting Brumaire, then the urban glamour of the fogs and mists that multiplied at the end of the nineteenth century – the Celtic twilight, the suffocating hothouse of *symbolisme* – were a more dangerous, because more alluring form of this fog. Another anathematizing of the atmospheric is to be found in the manifesto appended to the first number of Wyndham Lewis's *Blast!* in 1914. Before getting on to what we might regard as the serious acts of commination (Parisians, sport, Victorianism, the 'purgatory of Putney'), the manifesto blasts the English climate: [10]

BLAST First (from politeness) **ENGLAND**

CURSE ITS CLIMATE FOR ITS SINS AND INFECTIONS

DISMAL SYMBOL, SET round our bodies,
of effeminate lout within.

VICTORIAN VAMPIRE, the LONDON cloud sucks
the TOWN'S heart.

After attacks against the mildness of the Gulf Stream, which is responsible for various nauseating sins of mildness and temperateness in the English character, the manifesto returns to its atmospheric theme:[11]

CURSE

the flabby sky that can manufacture no snow, but can only drop the sea on us in a drizzle like a poem by Mr. Robert Bridges.

CURSE

the lazy air that cannot stiffen the back of the SERPENTINE, or put Aquatic steel half way down the MANCHESTER CANAL.

But ten years ago we saw distinctly both snow and ice here.

May some vulgarly inventive, but useful person, arise, and restore to us the necessary BLIZZARDS.

LET US ONCE MORE WEAR THE ERMINE OF THE NORTH.

If this form of modernism attempts to dispel the mists of glamour and of stupor, another had already begun to recognise in haze a particular comportment towards the in-between, to the background noise constituted by atmospherics. Increasingly, haze becomes a necessary condition of perception in the imaging of perception itself.

An important ingredient in the cultural atmospherics of the late nineteenth and early twentieth centuries was the haze of tobacco associated with esoteric and bohemian culture. Eugene Umberger has charted the large number of works in praise of tobacco and smoking that appeared in the late nineteenth century, such as Arthur Machen's *The Anatomy of Tobacco* (1884) and J. M. Barrie's *My Lady Nicotine* (1890), later subtitled *A Study in Smoke*.[12] The blue haze of the smoker's atmosphere was often presented as an externalized image of cerebration. For Sherlock Holmes, the solving of problems seems to require a more literal dissolution, in the conversion of large quantities of tobacco into smoke, a process that seems to transform the corporeal into the cerebral, as in this description of an all-night exercise in problem-solving from 'The Man With the Twisted Lip':

> [H]e constructed a sort of Eastern divan, upon which he perched himself cross-legged, with an ounce of shag tobacco and a box of matches laid out in front of him. In the dim light of the lamp I saw him sitting there, an old briar pipe between his lips, his eyes fixed vacantly upon the corner of the ceiling, the blue smoke curling up from him, silent, motionless, with the light shining upon his strong-set aquiline features. So he sat as I dropped off to sleep, and so he sat when a sudden ejaculation caused me to wake up, and I found the summer sun shining into the apartment. The pipe was still between his lips, the smoke still curled upward, and the room was full of a dense tobacco haze, but nothing remained of the heap of shag which I had seen upon the previous night.[13]

Smoking was also closely associated with psychoanalysis, and not only because of the well-known attachment of its founder to cigars. According to Lydia Marinelli, 'an unbreakable connection is posited between psychoanalysis, tobacco, and the disappearance of metaphysics' in Wilhelm Stekel's account of the early days of psychoanalysis in his 'Conversations on Smoking' of 1903.[14] Others, like T. S. Eliot, were less enamoured of the 'tobacco trance', and presented smoke as an image of a thought that had lost its grounds or definition, as in the early poem 'Interlude in a Bar': 'Across the room the shifting smoke/Settles around the forms that pass/Pass through or clog the brain'.[15] Another early poetic fragment by Eliot makes out from the inertia of smoke an image of matter sullenly persisting in defiance of human vigour:

> The smoke that gathers blue and sinks
> The torpid smoke of rich cigars
> The torpid after-dinner drinks
> The overpowering immense

After-dinner insolence
Of matter going 'by itself'
Existence just about to die
Stifled with glutinous liqueurs.[16]

One surprising philosophical resource for what we might call 'nebular modernism' is to be found in the work of Nietzsche. Surprising because, of course, Nietzsche, described aptly by Bachelard as the philosopher of ascension, also provides modernism with much of its rhetoric of eagle-eyed lucidity.[17] The alpine Übermensch looks down imperiously and disdainfully on the huddled, timorous, mist-saturated lowlands of cowardice and resentment. But, in *On The Uses and Disadvantages of History for Life* (1874), Nietzsche is to be found arguing for the necessity of a certain obnubilation. The historical sense is a 'vivid flash of light' (*ein heller, blitzender Lichtschein*) that breaks out of the 'encompassing cloud' (*umschliessenden Dunstwolke*)of unhistorical existence. Although, says Nietzsche, this makes a man out of a man, it carries severe penalties. For life requires forgetfulness, illusion:

> The unhistorical is like an atmosphere with which alone life can germinate and with the destruction of which it must vanish . . . When the historical senses reigns *without restraint*, and all its consequences are realized, it uproots the future because it destroys illusions and robs things of the atmosphere in which alone they can live . . . All living things require an atmosphere around them, a mysterious misty vapour; if they are deprived of this envelope, if a religion, an art, a genius is condemned to revolve as a star without atmosphere, we should no longer be surprised if they quickly wither and grow hard and unfruitful.[18]

Even more surprisingly, Nietzsche thinks that this atmosphere of illusion is so necessary and nutritive that it can actually help one to attain a 'suprahistorical' perspective: 'If . . . one could scent out and retrospectively breathe this unhistorical atmosphere with which every great historical event has taken place, he might, as a percipient being, raise himself to a *suprahistorical* vantage point'.[19]

J.M.W. Turner, *Sun Setting over a Lake*, 1840–45.

The affinity for mists and clouds in the late nineteenth century is represented in painting by the remarkable works of Monet and Whistler. Where the mist, fog and steam of many of Turner's canvases were dynamic, funnelling down into vortices of indistinctness, as though one were looking down from a thunderhead into a cyclone, the fogs of Monet and Whistler represent a kind of collision of the principles of incandescence and obscurity. This is a difficulty for the eye that is formed from dim dazzle, rather than from darkness. Here, light has become thickened into matter, as though captured like an insect in the slow ooze of resin. Haze is the complication of light: light made obscuringly visible. In haze, light is both magnified and congealed, in dense, yet ethereal detention.

These paintings have often been seen as harbingers of abstraction, since they seem to show the recession of their apparent objects away from the eye. We seem to be being taught to be content with the daubed approximations of the paint itself. It is unclear whether we are supposed to be trying to look through the pigment to what it seems both to present and obscure, or look at the obscuring effect itself. The medium of the painting is supposed to come between us and its spectral subject, even as it makes visible that betweenness.

Haze is particulate. The characteristics of haze, fog and mist derive from their scattering of light. Haze instances and anticipates what might be called the particulate dream of modernism, the dream of being able to register and merge with the infinite multiplicity of the atomic constitution of matter. Fidelity to what Woolf called a 'luminous halo, a semi-transparent envelope surrounding us from the beginning of consciousness to the end' is associated with the idea of 'an incessant shower of innumerable atoms; and as they fall, as they shape themselves into the life of Monday or Tuesday, the accent falls differently from of old'.[20] The cloud or the haze, 'vaguer in the air, more soluble' ('plus vague et plus soluble dans l'air') in Verlaine's phrase from his 1874 poem 'Art poétique', is the precipitated compromise between the conjoined but contradictory impulses towards the integral and the disintegrated to be found both in Woolf and in other modernists.[21] It is not clear what is the correct focal length for the viewing of such scenes, not clear where its viewer is supposed to be. As in pointillism, we can

Claude Monet, *Waterloo Bridge: Effect of Sunlight in the Fog*, 1903.

Claude Monet, *Charing Cross Bridge*, 1899–1901.

only see the painting when we can no longer see the elements of which it is composed. We can only see the painting when we have advanced close enough upon it for our vision to be decomposed or atomized.

If there is no privileged or prescribed place to occupy in order to make out the *Gestell* or disposition of the scene for the eye, similarly, there is no punctual moment of seeing. If mist or haze seems to suspend time, it also testifies to the fascinating, incessant drift of duration. Hence, perhaps, Monet's strange obsession with painting and repainting the same or similar scenes – the renderings of Charing Cross Bridge, or Waterloo Bridge, or the Houses of Parliament. They seem to suggest, not only a series of compositions, but also a composite viewing, an overlayering of filters, which themselves approximate more and more closely to the subject, which is obscurity itself. Michel Serres has often returned in his writings to the contrast between what he calls 'scenography' and 'ichnography'. In scenography, one sees singular appearances, made available at a particular moment, from a particular perspective. An ichnography is 'the ensemble of possible profiles, the sum of horizons. Ichnography is what is possible or knowable, or producible, it is the phenomenological well-spring, the pit. It is the complete chain of metamorphoses of the sea god Proteus, it is Proteus itself'.[22] This distinction arises from Serres's reflections on Balzac's story 'Le Chef d'oeuvre inconnu', which constitutes a large part of his book *Genesis*. At the climax of the story, the old painter Frenhofer shows his two friends his masterpiece, in which they can see '[a] mass of confused colours, hemmed by a multitude of bizarre lines forming a wall of paint'.[23] And yet he insists that the very dissolution of the subject is what he has sought: the essence of the woman he has painting is airy, cloudy:

> There is such depth in that canvas, its air is so true, that you cannot distinguish it from the air that is your element. Where is art? Lost, vanished! These are the very forms of a young girl. Have I not captured the colour, the life of the line that seems to complete the body? Is this not the same phenomenon as that presented to us by objects that are in their atmosphere like fish are in water?[24]

Serres takes this story, on which Henry James drew for 'The Figure in the Carpet', and which Picasso also greatly admired, as an allegory

Claude Monet, *Charing Cross Bridge, The Thames*, 1903.

Claude Monet, *Charing Cross Bridge, Fog on the Thames*, 1903.

for the idea of an art that would be able not only to redouble the visual information that perception picks out from background noise, but also the background noise itself, the formlessness that is a constitutive part of every perception, every sign:

> The work, through profiles, snapshots, Protean shapes, emerges from the perturbation, from the noisy turbulent sea around the island of Pharos, flashes, occultations, of the protophare. Without this pile-up, without this unknowable ichnography, there are no profiles, no work. It is necessary to dare to unveil the ichnography, at times, the one we always carry with us, in the dark, and as though secreted, in a receded nook, under a veil. Like a palette.[25]

One can find the same oscillation between radiance and fog, signal and noise, emergence and immersion, in the work of Joseph Conrad. Conrad, of course, insisted that the job of the writer is to 'make you *see*', but often what he wanted his reader to see was the effect of dimming and dazzling provided by haze.[26] The meaning of the seaman-narrator's yarns, Conrad informs us in a much-quoted and requoted passage, 'was not inside like a kernel but outside, enveloping the tale which brought it out only as a glow brings out a haze'.[27] This is usually read to mean that Marlow's stories, like Conrad's, have their meaning as a result of the lights cast across it, necessarily obliquely, by some exterior source of illumination – the listener, or reader, perhaps. As has often been noticed, one can track the appearances of haze and mist within Conrad's writing too – and Wendy B. Faris has drawn some extremely suggestive parallels between the rendering of haziness in the painting of Turner and Conrad's writing. Conrad's work oscillates between different meanings of the mist. Mist will sometimes represent a clarifying background, from which details may suddenly, precisely, be picked out, as in this passage from the beginning of *Heart of Darkness*:

> The day was ending in a serenity of still and exquisite brilliance. The water shone pacifically; the sky, without a speck, was a benign immensity of unstained light; the very mist on the Essex marshes was like a gauzy and radiant fabric, hung from the wooded rises inland, and draping the low shores in diaphanous folds.[28]

But, later on in the story, the white mist that descends on the boat that makes its way up the Congo river embodies disembodiment:

> When the sun rose there was a white fog, very warm and clammy, and more blinding than the night. It did not shift or drive; it was just there, standing all round you like something solid. At eight or nine, perhaps, it lifted as a shutter lifts. We had a glimpse of the towering multitude of trees, of the immense matted jungle, with the blazing little ball of the sun hanging over it – all perfectly still – and then the white shutter came down again, smoothly, as if sliding in greased grooves.[29]

The associations between photography and spiritualism at the end of the nineteenth century may be at work somewhere behind that image of the fog lifting and coming down again like a vast eyelid or the shutter of a camera. Photographers of mediums and seances not only sought to capture the spectral masses of spirit-bodies, or the billowing cumulus of ectoplasm, they also seemed to see an analogy between the actual apparatus of the photograph, so given to producing silvered mistiness, and this gelling of light or spirit-energy into indeterminate form. Conrad's interest in undulatory and radiation theories may also have helped him appreciate the effects of x-ray radiation, discovered only four years before the writing of *Heart of Darkness*. On the one hand, x-rays penetrated the flabbily obscuring veils of the flesh, to reveal the bony essence of what lay within. But x-rays also left visible traces of that pervaded flesh, dissolving it to a spectral haze or plasma, which, as Martine Hennard Dutheil de la Rochère suggests, seems close to the hollow, insubstantial bodies found in *Heart of Darkness* – the figure of Kurtz, for example, whose form will appear 'unsteady, long, pale, indistinct, like a vapour exhaled by the earth'.[30]

Rather than picking things out of the background, haze is a bringing forward of the background itself, as an infigurable, defiguring figure without ground:

> What we could see was just the steamer we were on, her outlines blurred as though she had been on the point of dissolving, and a misty strip of water, perhaps two feet broad, around her – and that was all. The rest of the world was nowhere, as far as our eyes and ears were concerned. Just nowhere. Gone, disappeared; swept off without leaving a whisper or a shadow behind.[31]

Interference

The change in signification of haze during the nineteenth century is accompanied and in some part enabled by the rapid development of atmospheric and meteorological sciences. One of the most important conduits of scientific ideas into literary and artistic life were the writings of John Tyndall. Tyndall had a particular interest in atmospheric phenomena, especially in relation to their optical effects. He took a strong interest in the question of how to make lighthouses effective in different atmospheric conditions, contributing a series of letters to *The Times* on the subject. But his interest in enhancing the clarity of visual signals cooperated with a highly developed sensitivity to the saturation of space – to its cooperations, compoundings and contaminations. For Tyndall, not just the physical atmosphere, but also the whole of cosmic space was continuously traversed, in all directions, by different kinds of impulse and radiation. As we saw in chapter Five, he proposed that the sum total of these radiations be called 'the temperature of space'. The atmosphere became 'a vehicle of universal intercommunication'.[32] Among Tyndall's more significant achievements was his explanation of the colouration of the sky. In later years, he became interested in atmospheric pollution, and an adherent of a theory of airborne infection that was tending to lose ground with the advent of the bacterial theory of disease. Tyndall was a significant contributor to the sense of the density and noisiness of space, permeated and perturbed by radiations of all kinds – of heat, light and electro-magnetic force.

What the atmospheric science of the nineteenth century began to display was that the atmosphere is not just affected by contamination and irregularity – it is constituted by it. The atmosphere is the ultimate 'mixed body', made of up of distributions, communications and interferences. In-betweenness becomes the normative condition of the atmospheric.

Tyndall's work would not have been possible without the confirmation of the wave theory of light at the beginning of the century, through the elegant experiment conducted by Thomas Young at the Royal Society that became known as the 'double-slit' experiment. This involved splitting a beam of light into two, and then allowing the two beams to come together again. Just as a wave that recoils upon itself when hitting an obstruction creates patterns of irregular amplification where peaks and troughs of waves coincide, so there were characteristic bars of lightness where periods of high light intensity coincided and patches of darkness where a peak and a trough coincided and cancelled each other out. As we have seen, the ether theories that flourished through the nineteenth century developed more and more complex accounts of physical phenomena, including matter itself, as due to the effect of various kinds of torsion or interference in the ether.

Haze is itself an interference phenomenon, in several senses. First of all, its optical effects come about because of the scattering of light rays by minute droplets of water suspended in the air. But it also embodies what might be called an interference of registers, a compounding of light and matter. Haze represents the interchange between the palpable and the impalpable, light made semi-solid.The population of the air with electronic impulses and radiations of all kinds vitiated the imaginary lucidity of the air, making for a new alertness to impediment and interference. Haze is a pervasive, versatile image of the signifying resistance – a resistance out of which signification comes – of the atmosphere. And, of course, interference became a new experience and a new metaphor for the unpredictable relations of things. If symbolism shifted into a poetic register the scientific apprehension of immateriality – the dissolution of solid matter into particles and forces in late nineteenth-century physics – then modernism began to see that a world of energies would be a world without permanent forms or distinctions.

Modernist haze brings the sky down to earth, or dissolves the grounds of the earth, dissolving the relations between sky and earth, creating interference patterns between high and low, frontality and immersion. The meaning of modernist haze is the loss of the sky – or, at least, the loss of its distance, its aura of unapproachability. More recently we have begun to see the emergence of an architecture of the mid-air. In 2002 the New York-based architectural firm Diller + Scofidio created an ephemeral structure as part of the Swiss EXPO 2000 in the waters of Lake Neuchâtel. Their Blur Pavilion was a suspended platform 92 metres long covered in an artificially produced cloud or fog, created from 31,400 jets spraying tiny drops of lake water into the air.[33]

Atmospherics

The specialized interest in the haze of modernism gives way to arts and protocols of the atmospheric, a generalized occupation of the spaces of traversal and passage provided by the air. The air that had previously been the outside or the stabilizing background of thought has become a populous in-between, a milieu of the mid-air. The calm, lucid infinitude of the sky has given way, as Richard Hamblyn, biographer of

Luke Howard, the inventor of the system of cloud-classifications used universally today, puts it, to 'skies vibrating, day and night, with an invisible topography of disturbance'.[34] As an example of the engagement of contemporary art, Hamblyn describes Usman Haque's remarkable sound installation of 2004, *Sky Ear*, which involved launching a cloud of 1,000 helium-filled balloons, each provided with a mobile phone, set to answer automatically. Once the balloons had ascended into the troposphere, the balloons could be dialled, and would answer with the electrosynaesthetic clamour of the middle air.

The work itself not only picks up interference, it is itself, like so much contemporary art, an effect of interference. As Haque explains, the original conception for the piece was simply to create airborne sensors with LEDs that would respond by changing colour as they encountered fluctuations in electromagnetic fields, thereby making the invisible visible. The effect of adding the mobile phones to the structure was to create the possibility of interfering with that process. The effect of listening to the sounds is actually to 'change the local electromagnetic topography and cause disturbances in the EMF inside the cloud that alters the glow, intensity and colour of that part of the balloon cloud'.[35]

But Usman Haque has also sought to imagine new forms of enclosure or privacy, amid this exposed, universally permeated condition. Another of his projects is to build a series of jellyfish-like structures he has called 'Floatables' that could provide ephemeral spaces of quarantine from data-exchange, in which individuals could take recuperative shelter. As far as I know, these floatables have not in fact been constructed. '*For many people the vessels are nothing more than a rumour.* Floating around urban environments, in the tradition of architecture that tries to break free from the confines of gravity, the vessels provide fleeting moments of private visual space, auditory space and olfactory space'.[36] Powered by sunlight and wind, the floatables fluctuate between the condition of actuality and dream, embodiments, perhaps of Woolf's 'luminous halo, a semitransparent envelope surrounding us from the beginning of consciousness to the end'.[37] They 'have no particular destinations and drift like flotsam around the city. However, they must keep moving because to be discovered by the authorities means almost certain destruction.'[38]

Romantic and nineteenth century writers were still able to keep the mists, fogs and hazes that were their subject in focus. For modernist artists, haze was part of the atmosphere of thought and perception, the very factor that inhibited artists from getting a fix on things. The enlivening that for modernist artists was how to write, paint, photograph, compose, from within the condition of the atmosphere. A century on the modernist topos of haze has become a topography, an environment, a locale, the hovering, indistinct space we are making our occupation.

There is another form of interference involved in the modernist atmosphere, namely that between eye and ear. Increasingly, the space of the sky is an aural space. Even its visible occupants and obstructions present themselves in terms of aural correlatives. Indeed, one might say that the concept of interference auralizes the space of the air. This kind of transcoding occurs in the white mist on the Congo river:

> I ordered the chain, which we had begun to heave in, to be paid out again. Before it stopped running with a muffled rattle, a cry, a very loud cry, as of infinite desolation, soared slowly in the opaque air. It ceased. A complaining clamour, modulated in savage discords, filled our ears. The sheer unexpectedness of it made my hair stir under my cap. I don't know how it struck the others: to me it seemed as though the mist itself had screamed, so suddenly, and apparently from all sides at once, did this tumultuous and mournful uproar arise.[39]

Atmospherics became the sphere in which a new conception of mixed and mutually pervasive bodies was worked out. The fortunes of the word 'atmosphere' itself express this. On the one hand, the word 'atmosphere' came to be used more and more to express the qualities of specific places or environments, according to the logic of the aura whereby a figure might be thought to exhale or extrude its own niche. But 'atmospherics' in general came more and more to mean the effects of interference, suggesting the confusions, interpenetrations, unpredictable mutations and compoundings of those places.

seven

Atmospherics

> What would the sun be itself, if it were a mere blank orb of fire that did not multiply its splendours through millions of rays refracted and reflected, or if its glory were not endlessly caught, splintered, and thrown back by atmospheric repercussions?[1]

Radio, or, more strictly because more broadly, wireless signalling, unleashed a dream of absolute communication and universal contact. Contemporary communications – or the material imagination that makes sense of them – still had as their ideal horizon a universe of absolute transparency and traversibility. In such a world, everywhere would be maximally accessible to everywhere else, and delay, obscurity and impediment would be done away with. It would be the opposite of Hobbes's vision of 'war of all against all': rather, it would be the communication of all with all. Atmospherics are the buzzing fly in that utopian ointment.

W. E. Ayrton's evocation of this world during a lecture given at the Imperial Institute in 1897 has frequently been quoted:

> there is no doubt that the day will come, maybe when you or I are forgotten, when copper wires, gutta percha covering, and iron sheathings will be relegated to the museum of antiquities. Then when a person wants to telegraph to a friend, he knows not where, he will call in an electromagnetic voice, which will be heard loud by him who has the electromagnetic ear, but will be silent to everyone else, he will call, 'Where are you?' and the reply will come loud to

> the man with the electromagnetic ear, 'I am at the bottom of the coal-mine, or crossing the Andes, or in the middle of the Pacific.' Or perhaps no voice will come at all, and he may then expect the friend is dead.[2]

There are two features of this that are worth comment. The first is the care taken to establish that the intercourse of the electromagnetic ear and voice takes place silently and secretly; the second is the total inundation notwithstanding of the air, earth and oceans. No matter where their adventures may take these wirelessly connected Jules Vernes, their existence will be so tied up with their ability to be in contact that failure to reply may safely be taken to mean death. We may borrow Hamm's words from Beckett's *Endgame*: 'Outside of here [the proleptic everyhere of universal hearsay] it's death.'[3] Some might imagine this situation of communicative incandescence as itself lethal.

This utopia of absolute communication is frequently evoked in the early years of radio. In 1912 *The Marconigraph* reprinted a poem from the *Melbourne Punch* celebrating the recent achievement of the SS *Miltiades*, which had sailed round the Cape to Melbourne without once being out of radio contact:

> There is no spot now where a man may go,
> Be it burning desert or Polar snow,
> But there the voice of a friend will come
> Bidding him hope for aid;
> There is no place now where the world is dumb,
> Or lonely, or left afraid;
> But universal are words like these,
> 'Hullo! "Miltiades!"'[4]

But this sociable ideal of general audibility encountered difficulties, which the following century would slowly come to suspect were intrinsic rather than accidental. The first is hinted at in Ayrton's evocation: that of secrecy. The necessity for some way to prevent spillage in a transmission system that, unchannelled by wires, would spread everywhere in all directions, was understood early, before the actual technology of radio had been developed. William Crookes saw in 1892 that signalling without wires through space would bring the possibility of eavesdropping and interference, though he was sadly optimistic about the possibility of keeping intruders out through tuning:

> I assume here that the progress of discovery would give instruments capable of adjustment by turning a screw or altering the length of a wire, so as to become receptive of wave-lengths of any preconcerted length. Thus, when adjusted to fifty yards, the transmitter might emit, and the receiver respond to, rays varying between forty-five and fifty-five yards, and be silent to all others. Considering that there would be the whole range of waves to choose from, varying from a few feet to several thousand miles, there would be sufficient secrecy; for curiosity the most inveterate would surely recoil from the task of passing in review all the millions of possible wave-lengths on the remote chance of ultimately hitting on the particular wave-length employed by his friends whose correspondence he wished to tap.[5]

Crookes does not anticipate a world in which electromagnetic ears will become capable of passing in review across all frequencies much more quickly than human ears. The difficulty of eluding other electromagnetic listeners was paralleled by the necessity of excluding unwanted or unexpected electromagnetic voices. For, towards the end of the nineteenth century, the air acquired a new accent. Users of telephones had for many years become inured to the annoyance of fizzing, crackling and other strange noises of electrical interference, and familiar with the haphazardness of 'bad lines' and 'good lines'. Even before the appearance of telephone wires, telegraph wires looped across the landscape from what are still in Britain called 'telegraph poles', seeming to suggest a kind of exposure to the air, and the possibility that the wires might leak sound, or the air might somehow become implicated in the messages it carried. But the development of radio, which would be identified with the air through which it was for the most part transmitted, rather than through the sea or earth, made for a new vulnerability of transmitted sound (and, later on, television images) to the vicissitudes of the air. Where previously the air had been audible only in the relatively familiar and recognizable forms of the soughings and screeches of wind, electrification gave the air a new, more diffuse, unpredictable and illegible sonority, a new, more enigmatic, more anguished music of the spheres. Put simply, the background of sound would come increasingly into the foreground in the same way as it did in an optical register in the case of haze. For that to happen, it was necessary that the background first be constituted as the channel of information – through the devising and diffusion of radiodiffusion. What came through *on* the air was the sound *of* the air, given voice by being given over to the electromagnetic carriage of voice. But, in the process, the air would become a different air.

This can be seen as part of a general widening of awareness within modernism to include the peripheral, the subliminal. Walter Benjamin called this new field of visual (in)attention 'unconscious optics':[6] we may similarly designate an 'auditory unconscious', constituted of everything that ordinarily fell upon the ear without being recognised or registered, but which nevertheless shaped feeling and perception. More and more, listening was to be assailed, augmented or interfered with by what made itself heard.

History of Atmospherics

The kinds of accidental interference that radiotelegraphers named, variously, 'strays', 'xs', 'atmospherics', 'parasitic signals', 'static' and 'sturbs' had a central place in the evolution of the theory and the material basis of radio. No sooner were radio waves detected and employed than the problem of atmospheric disturbance arose. It was clear early in the history of radio that atmospheric disturbances could produce and propagate the same kind of electromagnetic waves that Hertz had demonstrated by causing a spark to be transmitted across his laboratory.

But putting atmospherics back into the centre of the picture was a difficult, even a paradoxical enterprise because the point of understanding atmospherics was in order to suppress or expel them. Oliver Lodge, who in 1897 had taken out a patent on a tuning device that would enable radio to be transmitted and received without interference, declared bluntly that atmospherics 'are of no assistance, and are a nuisance which ought to be eliminated'.[7]

Indeed, atmospherics had come to notice even earlier than this. Even before the development of wireless telegraphy, telephone users found that their apparatus was subject to interference too, an interference that may itself have predicted some of the forms and uses of radio itself, though the effects were usually the result of electrical induction rather than electromagnetic radiation. One Charles Rathbone, who was listening on a private telephone run between his house in Albany and the Observatory, heard singing, which turned out to be emanating from an experimental concert transmitted by Thomas Edison over a telegraph wire between New York and Saratoga Springs. The *New York Times* carried a report in 1873 of the strange interferences produced in telegraph equipment by electric storms. The article explained that the 'electric wave' produced during a storm sometimes acted to block or obstruct transmissions, and sometimes augmented them. On occasion, it provided the possibility for a kind of wireless transmission of signals:

> When the electric wave is of considerable duration and power, the operators have been known to let go their batteries, detach the wires, carry them to the ground, and, by means of the electric throbs, messages have been transmitted entirely independent of the ordinary auxiliaries.[8]

The electrical nature of lightning had been known since Priestley, and suspected before, but it seems to have been Oliver Lodge who first proposed that lightning produced impulses of a specifically oscillatory character, just like the spark that Hertz used to show the existence of radio waves. With the telegraph or the telephone, reception and transmission had to be born together – that is, one could only a receive a message that has been humanly sent, despite the fantasies of spiritualists. Radio, by contrast, revealed an excited, excitable world of radio discharges, audible evidence of the universe of overlapping oscillations and radiations revealed by nineteenth-century physics. In the very earliest days of radio, one listened to, or listened out for, atmospheric impulses, since there was little else to listen to.

Research into atmospherics remained patchy and sporadic during the first decades of the twentieth century. The first systematic work was undertaken by H. Morris Airey and W. H. Eccles, who were able to observe in an important paper of 1911 that '[t]he sum total of the work published on the whole subject is very small'.[9] Slowly, as experiment and report began to build up, atmospherics started to gain a positive interest in themselves, rather than simply as a nuisance to be eradicated. A contributor to *Wireless World* wrote in 1920 that 'from the operator's point of view, these natural disturbances, called variously strays, or atmospherics, are particularly undesirable, though to the experimenter with non-utilitarian aims they present a fascinating field of study'.[10]

From 1910 onward, interest in and understanding of radio atmospherics increased markedly. The British Association set up a Committee for Radiotelegraphic Investigation in 1913, which undertook a systematic investigation of atmospherics. In 1918 Robert Watson-Watt began supervising research at the Aldershot Wireless Station. In 1920 the Committee for Radiotelegraphic Investigation gave way to the Radio Research Board, under the direction of the Admiral of the Fleet Henry Jackson. One of the four sub-committees it established was charged with the investigation of atmospherics.

A review of the subject in *Wireless World* in 1923 noted this huge increase in research, saying that '[n]ot very long ago it would have been easy to tell you in an hour's lecture – from a half-sheet of notepaper, so to speak – all that was known about atmospherics. Happily, that is not now the case.' Nevertheless, and despite the fact that, between 1906 and 1918, over a hundred patents had been lodged for anti-interference devices – or 'x-stoppers', as they were often known – it was still the case that 'the greatest unsolved problem in radiotelegraphy is that of interference by atmospherics'.[11] There were many false dawns. In 1919 Roy Weagant, a consultant with RCA, announced that he had discovered that the radio waves produced by atmospherics move at right angles to the waves produced by radio, and always in a vertical direction. This, he believed, would enable him to eliminate static altogether.[12] He was to spend four more years of largely fruitless research trying to use this insight to develop a foolproof means for filtering out interference. Despite the advances in understanding, there was no steady advance of clarity and corresponding retreat of atmospherics, partly because the advances in radio technology itself, such as the use of ever longer wave-lengths and more sensitive receivers, not to mention the huge growth in broadcasting, opened radio up to more sources of interference. A report on the Weagant 'x-stopper' in 1919 acknowledged that 'the interference due to these atmospheric impulses even in the old coherer days – bad though it then was – was almost as nothing compared with the terrific disturbances experienced in modern long-distance receiving stations'.[13]

Local Habitations

The first efforts at understanding atmospherics involved trying to identify their source. The assumption that guided most early research into atmospherics was that they were the result of lightning storms. In 1895 Alexander Popov connected a coherer to a lightning rod and showed that it was possible to detect approaching storms. Thereafter, others explored the possibility of weather forecasting by means of radio. In the early years of the century Albert Turpain made detailed observations that made it possible to detect thunderstorms many hours before there was any other evidence of them.[14]

W. H. Eccles and H. Morris Airey devised a system for recording 'strays', which involved making vertical lines of different lengths to mark different intensities of sound along a horizontal line representing the passing of time. When they correlated records taken at two receiving stations in Newcastle and London, they found a very close synchronicity between the two recordings, leading them to conclude that

> between 60 and 80 per cent of the atmospherics audible at Newcastle and London, about 270 miles apart, are due to the same cause. This cause is probably a discharge of atmospheric electricity at places whose distances from the stations are possibly of the order of hundreds of miles.

Eccles believed that most of the interference affecting British radio reception emanated from tropical storms in West Africa.[15] Others thought that audible atmospherics came from disturbances thousands of miles more distant still.

The study of atmospherics brought meteorology and radiotelegraphy close together, as is suggested by their conjuncture in the short-lived *Weather and Wireless Magazine*, which ran from 1923 to 1924. In its pages, J. Reginald Allinson concluded from the use of frame aerials to detect and track the progress of thunderstorms before there was visible evidence of them that 'these stray waves have been "captured," and made to serve a useful purpose'.[16]

But as atmospherics began to be more thoroughly investigated, they turned out to involve more than the weather. Radio emissions from volcanic eruptions suggested that radio would also have its uses for the geologist and the vulcanologist. A. G. McAdie wrote in 1913 that

> no great eruption can occur, with its development of atmospheric electricity, without a corresponding electro-magnetic wave disturbance in the ether, shown in the form of static interference on wireless receivers, and more or less pronounced interruption of communication by wireless telegraphy. The time is not far distant when by means of these various records it will be possible for the meteorologist to determine the velocity of propagation of volcanic ash, the detonation or sound waves, the pressure or impact waves, and the duration of thunderstorms and tornadic or whirlwind effects.[17]

Eccles devoted much attention to the mysterious fluctuations in the amount of atmospherics and in the range of radio transmissions between day and night. His explanation was that solar radiation must cause ionization of the upper atmosphere – in what was originally called 'the Heaviside layer', after the speculations of Oliver Heaviside concerning its existence, and then verified during the 1930s as the ionosphere – and that this ionization would be reversed during the hours of darkness.

Carl Størmer reported the puzzling phenomenon of echoes of short-wave signals being heard not just half a second after the source sound (the normal period taken for a signal to orbit the earth), but also at an interval of around 3 seconds afterwards. He speculated that the radio waves were reflected by the aurora borealis.[18] During this period some remarkable reports began to be received regarding the powers of the aurora borealis both to interfere with radio transmissions and, more remarkably, to become audible when the aurora occasionally came down to ground level. Writing in *Nature* in 1931, S. Chapman summarized the testimonies given by J. Halvor Johnson, who had investigated low-level auroral displays in Alaska and northern Canada:

> The sounds are variously described as 'a swishing or rustle like that of a silken skirt moving back and forth . . . very low, but yet plainly discernible'; like those 'that accompany small static discharges'; like the sound made when 'a couple of slices of good fat bacon are dropped into a red-hot pan'; 'they may attain a loudness comparable to that emitted by a high-tension electric current when charging a set of horn-gap lightning arresters'; 'quite audible swishing, crackling, rustling sounds'; 'a crackling so fine that it resembled a hiss'; sounds 'similar to escaping steam, or air escaping from a tire'; 'much like the swinging of an air hose with escaping air'; 'the noise of swishing similar to the lash of a whip being drawn through the air'; sounds 'likened to a flock of birds flying close to one's head'; 'not musical, it was a distinct tearing, ripping sound as when thin muslin is ripped or torn apart'. One man at sea, in an open boat with four natives, on Oct. 11, 1893, heard 'the most fearful whizzling and crackling sounds, sounding at times as if thousands of firearms were fired within short distance'; at the time there was 'no wind and no clouds'. Another writer mentions 'loud reports similar to rifle cracks', 'the air was still and the aurora was just above the tops of the birches'; the few loud reports were followed by much crackling.'[19]

Many of the radio waves emitted by the aurora occur in audible frequencies – which is not to say that they can on that account be heard, but they require only a transducer to pass across into sound, rather than any more complicated kinds of radio apparatus.

There were speculations about the possibility of picking up radio transmissions from outside the earth's atmosphere. Oliver Lodge had attempted in 1894 to detect radiation from the sun, but failed, as a result of mundane interference: 'There were evidently too many terrestrial sources of disturbance in a city like Liverpool to make the experiment feasible.'[20] During the 1920s, when the orbits of earth and Mars came close together, and following an injudicious hint dropped by Marconi, European and American newspapers became full of excited speculation about the possibility of picking up signals transmitted from the mysterious red planet. On 23 April 1920, Frederick Milliner and Harvey Gainer tuned in to very long wavelength transmissions in Omaha, in order to detect incoming signals from Mars. *The Times* reported the experiment rather coolly:

> At first (he said) we used wave lengths of from 15,000 to 18,000 metres, and for several hours it seemed as if we heard everything that was going on in the world. We got Berlin, Mexico, and all the large stations. We got in on a thunderstorm somewhere, and the crackling lightning was like hailstones on a tin roof all around us. About 2 a.m. it cleared up and everything grew quiet.
>
> Then we hitched up a long wave length, which took us into space – beyond everything that might be taking place on earth. There was a most deathly silence. We concentrated our faculties to catch the faintest sound, but there was nothing, nor was the silence broken during the entire time we had the long wave hooked on.[21]

This excitement was renewed in 1926, in which year *Popular Wireless* asked: 'Is it possible that the inhabitants of Mars will send a wireless message to the earth on October the 27th? On that date, the mystery planet makes its nearest approach to earth. Can the vast space separating planet from planet be bridged by radio?'[22] *Popular Wireless* put together a powerful fourteen-valve receiver, and claimed on 6 November 1926 to have received a mysterious signal: 'Who sent the mysterious M's that were picked up on the PW 14-valve set, when listening-in for Mars? . . . several expert telegraphists were among the company that actually heard the M's, and there is no doubt whatever of their mysterious nature.'[23] The technical press was rather sniffy about this popular effervescence. In 1920 *Wireless World* published a picture of a Mr Frank Marshall receiving signals in the cellar of the Rose and Crown in Park Lane, which bore the sardonic caption 'He is NOT receiving from Mars.'[24]

This worldliness was defeated in 1931 when Karl Jansky, investigating the problem of atmospherics in transatlantic communications for the Bell Company, discovered that, even when one subtracted the static produced by known atmospheric disturbances, such as thunderstorms, a residual noise persisted, which underwent a periodic variation corresponding to the period of the earth's rotation with respect to the stars.[25] When Grote Reber built his own bowl-aerial that he could point to different areas of the sky, he found that the radio emissions were strongest from the parts of the Milky Way in which stars were clustered.[26] The discovery of this 'cosmic static' would lead after the Second World War to the huge advances in the understanding of the universe brought by radio astronomy.

In one sense, the mapping of radio space has helped to put and keep atmospherics in their place. The audible atmosphere was an atmosphere that lost its traditional dimension of altitude: hereafter, one might be airborne, or in mid-air, communicating with the air by communicating through it, without actually having to be aloft. A celebration of the life of Marconi in the inaugurating number of *Wireless World* said that 'he found the bridle which controlled this Pegasus of the air, and as a result our second Prometheus brought down to earth "radiotelegraphy"'.[27] But, as radio has been steadily spatialised, so space has been radiolized – that is to say, reconfigured to accord with a world in which what matters are not points, nodes, orientations and distances, but velocities, frequencies, connections, transmissions and syntonies. Radio space could come as close to the surface of earth as the ground-level aurora borealis, and extend further into space than optical telescopes could reach. Radio allowed the experience of the far-here, or the far-hear: that which was unimaginably distant could also have the immediacy and importunacy of that which sounded in your ears.

And a Name

One of the odd symptoms of early attention to atmospherics was a desire to formalize them in a descriptive and expressive language. The noises that interrupted and sometimes swamped communications were not merely random, but had their own acoustic profiles. A phonology and then a phonetics of the atmosphere began to be devised, as the mouth and tongue assisted the ear in picking out, naming and echoing back in language the different kinds of interference. Observing that the electro-magnetic atmosphere 'had a language of its own', J. J. Fahie wrote that the sound of lightning discharges registering on telephone lines was 'very characteristic – something like the quenching of a drop of molten metal in water, or the sound of a distant rocket'.[28] As early as 1913 Eccles proposed a distribution of radio atmospherics into 'clicks', 'grinders' and 'hisses' (or 'fizzles'). A writer for the journal *Wireless World* in 1919 explained that '[h]issing noises are due to actual static discharges from the aerial to earth caused by electrostatic induction by charged clouds or winds', while clicks and grinders emanated from lightning discharges, and were most common in transatlantic communication.[29] An article in *The Times* in 1925 added a couple more terms to this taxonomy of aerial tinnitus:

> crashes . . . which may last as long as five seconds and appear to result from local temperature changes and squally weather',

and the 'fizzly',

> quite a distinct kind of atmospheric which often accompanies rain and hail squalls. It causes a continuous hissing sound in the receiver, and only occurs when showers of rain or hail having charged particles are near or actually in contact with the aerial.[30]

Some researchers developed specialities among the forms of atmospheric disturbance. Heinrich Barkhausen reported in 1919 on his work on long whistling tones. He explained that, during the years when military radio operators employed enhanced amplifiers in order to try to intercept transmissions from the enemy, they would regularly hear on their headphones

> a quite remarkable whistling tone. At the front, it was said that one heard 'the grenades flying'. As far as it is possible to represent it in letters, the tone sounded somewhat like *piou* . . . beginning with the highest audible tones and then running through the whole scale of tones and finishing with the lowest audible tones. On many days these whistling tones were so strong and frequent that they sometimes made listening in impossible.[31]

Barkhausen originally thought that these tones must originate from the earth, since the radio apparatus on which they were detected was often deep underground, but, when he returned to the subject a decade or so later, he revised this opinion, ascribing the whistler to multiple reflections from the ionosphere.[32]

The coming of broadcasting in 1922, and the subsequent crowding of the airwaves, meant that many more listeners became familiar with the effects of interference, and new sounds began to be distinguished. A feature article in *The Times* described 'radiation from oscillating aerials, which is generally known as "howling"' and also two kinds of interference from Morse signals – '"mush", a welter of vague unreadable dots and dashes, and rustling noises, whistles, and chirps, produced by the harmonics of continuous-wave stations'.[33] Specialized terms like 'heterodyning', to describe the effects of combined frequencies of different kinds of receivers, started to pass into ordinary use. Describing observations and experiments with different kinds of interference picked up by submarine cables, E. T. Burton and E. M. Boardman carefully distinguished two new varieties of musical atmospherics – the 'tweek', a 'damped oscillation trailing a static impulse' and the 'swish', which has a sound such as is made 'by thin whips when lashed through the air'.[34]

More recently, groups of radio researchers have taken to listening out for VLF or Very Low Frequency signals from nature, the particular fascination of such materials being that, like the ground-level aurora borealis described earlier, they are often in the audio range of frequencies. This has spawned an exotic zoology of sonorous sub-species, including 'hooks', 'risers', 'pure-note whistlers', '2-hop whistlers', 'whistler echo-chains' and 'dawn-choruses'. Perhaps the suggestion here is that atmospherics enact a kind of incipient self-naming, seeming almost to speak themselves, as though a voice were emerging out of the fog of noise (as a *noice*, or a *voise*).

Tapping, Tuning, Jamming

But, increasingly, there was another, endogenous form of interference that had a less celestial origin, and came from the inside of radio communications. The multiplication of different kinds of electrical appliance, including radio appliances themselves, brought about forms of human atmospherics. The 'Amateur Notes' column of *Wireless World* in 1913 sardonically reported the concern of established radio users at 'interference caused by the learners at these schools [wireless schools in London] transmitting too diligently' – and, what is more, using magnetic receivers.

> We rather gather that the frame of mind of those correspondents who are protesting against the conduct of these schools is this: 'If I, by careful adjustment, can receive signals on my chalcopyrite-molybdenite crystal – quite five times out of ten – without missing more than a few words now and then when my crystal goes out of adjustment, where should there be any need for anyone else to use the magnetic detector, which is undoubtedly less sensitive, and therefore requires stronger signals which interfere with my experimenting?'[35]

As the airwaves began to become congested, the human atmosphere started to provide new sources of involuntary or accidental interference. C. G. Blake complained that 'I have very great difficulty in my own station, because there is a butcher's shop very close where they work a sausage machine, and when it is going it is absolutely impossible to receive signals at all.'[36] 'B.C.L.' wrote from Colchester to *The Times* with his speculations about the origin of the persistent atmospherics that affected radio reception in snowy or frosty weather, suggesting that they were caused when the wheels or connectors of local trams no longer made good contact with cables or rails, and therefore sparked.[37]

Neighbours using different kinds of equipment could easily interfere with each other's reception, as *The Times* carefully explained in 1926:

> The crystal set is regarded by many as being entirely incapable of giving rise to any kind of interfering noises. Actually a crystal set may be even more annoying to a neighbour than a valve receiver, should the respective aerials be close together and run parallel with one another; for whilst one is searching for a sensitive spot with the point of a catwhisker, one's neighbour may be pulling his set to pieces in the hope of discovering the cause of a baffling series of crackles, crashes, and grinding noises. The two aerials are tuned to the same wavelength. When the catwhisker is raised from the crystal the path to earth from the aerial of the set of which it forms part is broken, to be made again directly contact is re-established. As the catwhisker is moved over the surface of the crystal contacts of varying resistance are made. Once again the carrier wave is slightly modulated, with the result that the valve user next door hears a succession of weird parasitic sounds. If when the sensitive spot has at last been found the crystal user lays his telephones on the table for a moment and gives verbal expression to his joy, it is quite possible for his words to be heard by his neighbour, since the diaphragms of the telephones vibrate under the influence of the sound waves and the carrier-wave is once more modulated.[38]

Sometimes, the human atmosphere seemed to be displacing the natural. Paul Augsburg published in 1927 a collection of stories on the theme of radio in which he evoked the human turbulence encountered by the northwest wind:

> [N]ow even a plain nor'wester can't howl a bit without getting tangled in a most amazing assortment of saxophone blues, stock quotations, tenor grace notes, hints on how to hold a husband, and what to do when your partner bids three hearts – all pushing relentlessly to keep a rendezvous with the peepul.
>
> Nor are these the sum of strange things that the plain nor'wester encounters. Sometimes it runs into a play being broadcast over the radio. A voice cries, 'Stand back, you bully!' and the nor'wester, amazed, asks, 'Who-oo-ooo?'[39]

Not all of these forms of interference were accidental. There are two kinds of distortion possible of the perfect transmission. The first is an implied diminution of the signal, through appropriation or tapping. This may make the signal intelligible, but insecurely transmitted. The second is an unwanted augmentation of the signal, by extra sounds that may make the signal fully transmissible but imperfectly intelligible. These alternatives of diminution and augmentation quickly became known as tapping and jamming.

Jamming began early in the history of radio, especially where rival systems or commercial interests were at work. The first congestion was experienced at sea, and this contemporary account describes some of the effects of interference and means used to combat it:

> The spark sets of those days – 'rock-crushers,' they were humorously called because of the deafening noise they made – had so broad a band of action and made so terrific a clamor that two fellows chatting thus practically blanketed any other vessel within fifty miles that might want to use the air.
>
> The only way to choke them off – and it was frequently used by some other operator waiting for a chance to send – was to 'drop a book on the key'; that is, lay a book or some other weight on the transmitting key, setting up such a continuous roar of interference that nobody within range could send or hear a word – rendering confusion worse confounded. These and the many other inevitable interferences when everybody was operating at will on the same wave length naturally led to wireless quarrels

> and feuds, filling the air at times with curses, aspersions and choice obscenities.[40]

One of the earliest and most highly publicized episodes of deliberate jamming occurred in 1903, during a demonstration of Marconi's system of transmission at the Royal Institution. Nevil Maskelyne set out to show that Marconi's system of tuning or syntony was neither as secure nor as immune from interference as he claimed. Using an untuned transmitter at the Egyptian Theatre, which transmitted 'dirty waves' across a wide spectrum, Maskelyne transmitted the word 'rats' repeatedly to the technicians who had their equipment set up to show the Morse receptions on the platform, followed by a limerick beginning 'There was a young fellow of Italy/Who diddled the public quite prettily.'[41] J. A. Fleming, who gave the lecture, wrote to Marconi the following day, alluding to the 'dastardly attempt to jamb [*sic*] us', and then, still steaming a few days later, to *The Times* to protest at this 'scientific hooliganism'.[42] This brought a defiantly self-justifying reply from Maskelyne, who claimed that he had undertaken the exercise as a scientific demonstration: 'We have been led to believe that Marconi messages are proof against interference . . . But when we come to actual fact, we find that a simple untuned radiator upsets the "tuned" Marconi radiators'.[43]

Such episodes became more and more common. When, in 1904, the US Navy set up tests of an interference preventer that had been devised by Reginald Fessenden, he discovered that the rival De Forest Wireless Company had hired an operator to disrupt its transmissions. The ensuing events are described by a witness:

> In an endeavour to hold off this interference the DeForest operator was kept under the influence of strong liquids during the tests, but in an unguarded moment he slipped away from his guards, got back into the radio station and started up a powerful transmitter, placing a brick on the key. In a few moments there was a knock at the door of the Navy Wireless station and there appeared Mr Dan Collins, the DeForest operator, who demanded that food and drink be supplied forthwith, or he would refuse to take the brick

> off the key . . . it was not till Mr Collins was finally induced, thru the supply of food and drink to withdraw the brick, that the Navy tests proceeded.[44]

The First World War saw systematic attempts at jamming and intercepting transmissions. One of the earliest appearances in fiction of the word 'jamming' to mean deliberate blocking of wireless signals occurs in 'The Vindication of Binsted, Ex-PO, a sea story by Patrick Vaux of 1914. The story deals with the detection of a spy who has stolen information that enables him to jam the signals of the Navy in the North Sea from an airship.

> 'Humberstone waves blocked again,' said the C.O. in surprise to Kelsale as he stepped from starboard on his bleak, high bridge, where the molecules of fog bleared the eye and made everything wet and clammy. 'Oh, damn this fog coming down again. Ouch! We'll stop this jamming, wherever it's coming from. It almost seems as if some folks have got inside our new transmitters.'[45]

Here, the word clearly refers to the efforts of unknown human agencies, but the association of this with the clammily rhyming conditions of fog, which seem to exemplify the state of radio isolation, perhaps suggests an interference of means. For indeed, the transitive use of the word 'jamming' to signify the deliberate blocking or confusion of signals by human interference coexisted until well into the 1920s and perhaps beyond, with another usage that referred to accidental atmospherics. W. E. Collinson's record of his own language use includes 'jamming' among the words most widespread among non-technical people in 1925: most such people, he writes, 'will have some inkling of the mysteries of *tuning in* and *tuning out* and the trouble caused by *jams, atmospherics* and *howlings*'.[46] Here the word 'jamming' means 'becoming jammed', rather than the action of jamming another transmission. When Reginald Allinson referred in 1924 to the efforts being made to rid wireless reception of 'Nature's jammings', the phrase seems nicely poised between accident and intent.[47] The use of the term 'jamming' in jazz, to mean spontaneous improvisation of two or more musicians together, seems to have been established by the early 1930s, and perhaps before. It is one of the many ways in which the idea of atmospherics begins to move from the distorting outside of music to its inside.

Probably the narrowing or focusing of the meaning of jamming came about after the Second World War. This was the first major conflict in which radio was not a mere psychological or technical accessory, and in which not only did radio become a means of waging warfare, but also sound and the study of conditions of transmission, audibility and intelligibility became an important area of research, as attested to by a review article on wartime research that appeared in 1948. The 135 items in the bibliography appended to the article include studies with titles such as *The Design of Jamming Signals for Use Against Voice Communications*, *Speech Transmission Through Six Military Gas Masks*, *Experiments With Earplugs: Their Effect on the Intelligibility of Speech*, *A Modified Tank Crash-Helmet for Use With a Separate Telephone Headset*, *Physiological Effects of Exposure to Certain Sounds*, *Methods of Training Telephone Talkers*, *Speech in Noise: A Study of the Factors Determining Its Intelligibility*, *Transmission and Reception of Sound Under Combat Conditions*, *The Effects of Noise and Vibration on Psychomotor Efficiency*, *Effects of High Altitude on the Human Voice*, *The Development of Ear Wardens*, *An Electronic Device to Simulate Atmospheric Static*.[48] It was out of this general effort to distinguish the conditions under which signals could be distinguishable from noise that Claude Shannon's mathematical theory of information would arise.

The great generative problem for twentieth-century communications, which is repeated in accelerated forms in the contemporary race to extend bandwidth and computing speeds, is the problem of how both to propagate and to regulate the space of communications. Every attempt to extend range, power and sensitivity – to open up new radio space – brought with it the possibility of new forms of interference. Every attempt to extend, diffuse and amplify the body beyond its limits brought exposure to the corrupting or complicating body of noise in the channel. As Hugh G. J. Aitken has argued, the rhythm of radio, that great rider of the vehicle of oscillating impulses, is itself an oscillation between the opening up of a new dimension, 'whose nature and dimensions could be grasped only by the scientifically trained intellect, one in which there were no familiar landmarks or units of measurement, one where place, occupancy, and possession had to be given meanings different from any they had had before', and crises of overcrowding and trespass in the new 'electromagnetic continent' that had to be met by struggle or international regulation – or large-scale interference and tuning effects.[49]

The Time of the Air

In his 1933 manifesto 'La Radia', Filippo Marinetti predicted that radio would not only do away with distance, it would also abolish temporal divisions, as the whole world participated in a perpetual, dynamic 'now' of radio-time, 'without time or space without yesterday or tomorrow'.[50] The movements of the atmosphere give rise to the weather that is time. In a certain sense, the atmosphere is time itself. As we have seen, early measurements suggested that the atmosphere was an arena of previously undreamt-of instantaneity. We should bear in mind that one of the most important early uses of wireless telegraphy was to broadcast accurate time signals, such as those that were transmitted from the Eiffel Tower. In response to a lecture on the history of the chronometer in 1920 a discussant identified only as 'the Hydrographer' pointed out that

> It is possible now, each day, for all those who have these instruments and receiving apparatus and are within range – by means of the Vernier time signals sent out from the Eiffel Tower – to accurately determine errors easily within one-hundredth second of time, by this delicate method, which I have no doubt many here are familiar with. By means of a chronograph and a relay you can reduce that error to something very much less, and the difficulties of interference such as atmospherics and jamming are largely surmounted.[51]

This seems odd – though it is easy to understand how atmospherics could obscure a time signal, the suggestion here seems to be that atmospherics and jamming might themselves introduce a kind of temporal error. If radio could be used to synchronize the world, it was also subject to more unpredictable fluctuations in electromagnetic weather. W. H. Eccles led the way in investigating the patterns of intensification and abatement in atmospherics. He was curious about why radio reception was better at night than during the day, and why atmospherics should exhibit the kind of fluctuations he described in 1909:

> Starting to listen at about a quarter of an hour before sunset the strays heard in the telephone are few and feeble, as they have been all day. Then, at five minutes after sunset, a change sets in, the strays slowly get rather fewer and feebler, till at 10 minutes after sunset a sudden distinct lull occurs and lasts perhaps a minute. Occasionally at this period there is a complete and impressive silence. Then the strays begin to come again. They quickly gain in number and force, and in the course of a few minutes they settle down into the steady stream of strong strays proper to the night.[52]

Part of Eccles's explanation was that when the rotation of the earth carried a portion of the atmosphere out of the sunlight, it formed a region in which the ions caused by solar radiation began to recombine. Endlessly revolving round the globe, this twilight belt was 'the seat of perpetual electrical discharges'.[53]

What was heard in atmospherics was the fracture and fluctuation of time; atmospherics suggested a time out of joint. As broadcast radio became more established, atmospherics often took the form of an infuriating fringe of Morse signals, which seemed more and more to intervene, not just from a different frequency, but also from a different epoch of communicative time.

Recalling the nineteenth-century belief that the ether was a kind of limbo or aerial graveyard of lost sounds, an article entitled 'The Music of the Spheres' in the *Daily Chronicle* of 10 December 1927 proposed that 'the eloquence of Cicero and Demosthenes may be recaptured for all to hear'.[54] In his 'A Remembered Queen', Siegfried Sassoon tuned in to the common fantasy that the ether, and especially the regions of it inhabited by radio static, might provide a carrier wave for the voices of the long-dead – in his particular case, the person of the twelfth-century 'wild and warring' Queen Mathilda:

> Did voices walk the air, released from death,
> Hers might be heard when, very late at night,
> I turn the wireless on and catch no sound
> But atmospheric cracklings, moans, and thuds.
> Hers might be heard, associate with this ground
> Whereon her house once stood.[55]

Sassoon picks up the language of picking up and tuning in, offering a strangely apt prepositional invention in his use of the phrase 'on the dark', in place of the more familiar or expected 'in' or 'through', and therefore suggesting that the dark was not so much a medium as a programme or broadcasting station:

> . . . If on the dark
> I heard shrill Norman French and stood between
> That utterance and eternity! If, so
> Attuned, I could watch Queen Matilda go
> Hunched on her horse across the crunching snow![56]

Breakthrough

It could only be a matter of time before spiritualists and supernaturalists, who had kept a close eye on electronic communications from the beginning, would see the opportunity that lay in atmospherics. In the 1950s Friedrich Jürgenson thought he heard anomalous voices on recordings of birdsong. In 1964 the Latvian parapsychologist Konstantin Raudive read of Jürgenson's claims and began working with him to try to detect and record Electronic Voice Phenomena, the voices of the dead, often by tuning a radio to the static between broadcast frequencies, or recording from an untuned diode. The voices, who were often those of recognizable media celebrities, spoke in a polyglot jabberwocky, their preferred mode of address being the gnomic yelp or bark: 'Mark you make believe my dear yes', was Winston Churchill's puzzling admonition. The breakthrough, and the breaking through of the voices, was announced in a book of that name in 1971.[57]

The atmosphere has often made its moods and powers known through sound – in wind, rain and thunderclap. But the sound of the atmosphere is no mere accompaniment or correlative, in the manner of an animal's cry. The atmosphere sounds in the mode of irruption, or intemperate exclamation, as though in agony, prodigy or augury. Something that is ordinarily mute and impersonal suddenly seems to 'sound', to give voice – or be given to it. The voice of the atmosphere is not something that it has as a resource or property, lying, as it were, ready to hand for it. For the atmosphere to sound is for it *to break into* sound, to pass across some inhibiting boundary, overcome some blockage.

Radio represented the universal trumping of gas by electricity. Gas was organic, approximate and aleatory. Electricity was instant, mathematical, absolute. Radio waves were thought of as penetrating. The atmosphere constituted a distance to be traversed or an obstruction to be pierced. In July 1909 *Technical World Magazine* carried a report about the work of a French inventor named Maurice Dibos, who claimed to have invented a device that would clear the air of fog by broadcasting radio waves in combination with hot air.[58] Dibos may have known of Oliver Lodge's early work with electrostatic precipitation, which showed that electrical fields could clear regions of the air of particles of matter, which could be charged and then attracted to plates from which they could be mechanically cleared. Lodge had experimented with fog dispersal aerials in Birmingham in 1903, and his work was sufficiently advanced for him to form the Lodge Fume Deposit Co. Ltd in 1913.[59]

The atmosphere emerges as a medium to be traversed and therefore a resistance to be overcome. But the obstruction of the atmosphere is not like obstructions to light, which result in a visible subtraction or diminution. The absence of sound is not a dimming but an augmentation and a complication. The shadow that falls across radio-sound is made of radio-sound itself. Because the nuisance will always assume the form of a noise, the accidents and infiltrations of the atmosphere can always come to constitute the signal.

When the atmosphere itself broke into radio, it was an inversion of the usual pattern. The obstruction had penetrated the penetrating medium. In its mildest forms, atmospheric sound is mere distraction or nuisance; in its major forms, it is often associated with crisis, catastrophe, even annunciation. The event of sound is never a completed arrival or coming to rest. Sound is never complete or definitive. Sound is always interrogative, asking, or prompting the question – 'what am I?' No serious examination of atmospherics could be done by the ear alone, since the ear never captures aural events; rather it is exposed to and taken by them.

And yet there was a strong impulse to bring atmospherics into a condition of what might be called aural presence, as though to hear these phenomena was to be in their vicinity in a way that transcribing their traces could never provide. It was not that one could wholly capture these sounds: but one could organize and orchestrate one's exposure to them. Thomas Edison's assistant A. L. Kennelly wrote in a letter of 2 November 1890 to Professor Holden, principal of Lick Observatory, of their ideas for detecting the electromagnetic radiation of the sun: 'Along with the magnetic disturbances we receive from the sun which, of course, you know we recognise as light and heat . . . it is not unreasonable to suppose that there will be disturbances of much greater wavelength. If so, we might translate them into sound'.[60] (But why?)

Edison and Kennelly anticipate more recent attempts to realize electromagnetic perturbations as sound. Perhaps there is an implicit understanding here that sound is always a matter of what breaks out from a background. This anticipates radio astronomy, which, far from allowing us to hear the eternal and unchanging music of spheres, gives us the sounds of violent catastrophe. As F. Graham Smith points out,

> Radio waves pick out the variable objects; these are often exploding violently, sometimes as individual stars but more usually as the central cores of galaxies. This violence manifests itself in radio emission, since radio comes naturally from the hot ionized gas surrounding exploding objects; furthermore this gas, and the radio waves it emits, can change much faster than the more condensed

> objects which radiate most of the starlight. The radio sun is extremely variable: at times it can burst out with dazzling brightness on the longer radio wavelengths.[61]

There were benign forms of this breakthrough. Since sound was thought of as a kind of spontaneous spilling or overflow, it could also be the warrant of a personal presence, a kind of authenticating parasite. Radio would supply those tones of voice that were seemingly not available in the encoded forms of signalling that characterized the telegraph. But even the telegraph had developed a complex kind of interference of its own, an interference that did not inundate or impede the message, but was rather conveyed with it, as a timbral aura or noisy 'voice', giving expressiveness and character even to Morse transmissions. 'Nothing could be simpler than its alphabet of dots and dashes', wrote L. C. Hall, in 1902,

> Yet it has come to pass that out of the manner of rendering this simple code has been evolved a means of communicating thought and feeling rivaling in flexibility and scope the human voice . . . A telegrapher's Morse, then, is as distinctive as his face, his tones, or his handwriting; and as difficult to counterfeit as his voice or writing.[62]

Hall tells a story of an acquaintance of his, whom he had known only by his call-sign of C. G., but whose character and disposition had been conveyed in the manner of his Morse. He reports that, when C. G. lay dying in a hospital, all his efforts were to get a message through to Hall. The alternation between internal speech and attempted communication was expressed in an alternation of different voices: 'While he tapped out his messages he spoke in a tense half whisper, like one trying to project his voice through space. Between times, however, in communing with himself, he spoke in his natural tones'.[63]

The indefinable personal qualities that are conveyed in the rhythms of the Morse are also themselves subject to noise and obnubilation. The accepted patois and abbreviations of Morse, directly equivalent to the 'txt-talk' of contemporary text-messaging and chat-lines, could throw up interference if they were transcribed exactly as they sounded, especially in the kind of 'hog-Morse' transmitted by an inexpert or clumsy operator. The literal sound of the Morse could also generate interference, an interference that then becomes the characteristic sign of certain personalities:

> The mere sound of the styles of some transmitters is irresistibly comic. One of these natural humorists may be transmitting nothing more than a string of figures, and still make you chuckle at the grotesqueness of his Morse. It is an every-day thing to hear senders characterized as Miss Nancys, rattle-brains, swell-heads, or cranks, or 'jays,' simply because the sound of their dots and dashes suggests the epithets.[64]

The Work of Listening

The breakthroughs of atmospherics consistently construe listening and the forming of sound as an ordeal – in strong contrast to the evocations of effortless overcoming of distance. Early radio constituted an arduous, attentive, inventive labour of listening. If atmospherics forced themselves into audibility, then the early radio listener was all the time exerting his own reciprocal pressure to grasp at the fugitive sounds he sought. Listening was a work of eye and hands as well as of the ear, in which there is not much that looks or sounds like passive 'reception'. The work of listening was an active interception. The hams, amateurs and hobbyists who took radio forward in the teens of the century, filling the technical journals with the excited buzz of their witness, discovery and speculation, constituted a laborious manual imagining of the immateriality of a culture constituted of wave-forms. Radio equipment was cumbrous, mysterious, frustrating, fascinating, with the coils, wires, batteries, crystals, chokes, condensers, capacitors, coherers, diodes and triodes, and all the manifold ways in which they could be configured and conjugated. Apparatus had to be designed, assembled, tested, adjusted, reworked, for radio to be developed. Listening in was itself a kind of R and D. Later, as radio transmission was taken over by official agencies, especially during the First World War the armed forces, and, from the 1920s onwards, broadcasting companies, amateurs came to constitute a kind of fringe phenomenon, a form of interference themselves.

Just as the transformation of the phonograph into the gramophone turned what had been a kind of alternating current, in which production and reception repeatedly changed places, into a direct current, in which the production was concentrated at one end of the process, and reception at the listening (and purchasing) end, so the roles of transmission and reception were increasingly polarized in the experience of radio.

This resulted, and was expressed in, some striking modifications of receiving apparatus. Where early radio apparatus had been ungainly and distributed, involving a number of different components, radio sets from the 1920s onwards tended to take a more and more integrated form. Radio apparatus came indoors from the sheds, basements and workshops, into living rooms, where the radio set was disguised as furniture. The word 'set' changed its meaning – no longer a collection of components that needed to be carefully set up, the set had settled into a single, bounded form. Much of this work of removing the work of listening involved the simplification of tuning. Tuning to a particular wavelength is achieved by varying either or both of the inductance or capacitance of a circuit. Tuning in early wireless set-ups was a complex and delicate affair, usually requiring three separate knobs to be turned until a signal or station was heard. The station would then be fixed with small adjustments of all three knobs, and their positions recorded – though a change in the position of aerials or the replacement of components could mean that that the settings would have to be recalibrated. The first device to allow tuning with a single switch or knob was patented in 1925, and thereafter was quickly established.[65] Pre-tuned push button controls appeared in the 1930s. The control of the wayward emissions of radio led to the development of radio as a method for remote control. At the Radiolympia show in 1933, Marconi exhibited a radio that would automatically tune into a desired station when its name was called out.[66]

The defeat of interference and the growing regulation and regularization of the airwaves was accompanied and symbolized by the appearance of the tuning dial. Although newspapers and specialist radio magazines carried details of the location of stations and their schedules during the 1920s, there were usually no names of radio stations inscribed on the apparatus itself; instead, there was usually a strip calibrated at 100-metre wavelength intervals. After the international agreements of the early 1930s had more or less stabilized the positions of different stations, dials began to take a circular form, with stations arranged in a kind of zodiac, as in the famous Ekco AD-65.[67]

Where previously the radio listener was in the midst of the circuit, radio developed a face, and listening began to be a face-to-face activity, with the listeners placed in front of the source of sound, a principle that seems to have been grasped early on in the design of the German 'People's Radio', or *Volksempfänger* VE 301, of 1933. This too reinforced the sense that radio had a finite itinerary, with the broadcasting source at one end of the process, and the listener at the other end, as the destination of the sound. This was the period in which listeners clustered round the radio to look at it, as though radio were already turning into television.

All this time, atmospherics were apparently being slowly squeezed out of the system of radio, which became ever cleaner, more efficient and more tuned. But atmospherics did not thereby simply recede from notice. Instead, having been removed from the medium, atmospherics returned as part of the message: atmosphere came into the foreground. One way of defeating atmospherics was to exclude them from awareness: the other was to bring them to attention.

Atmospherics began to be thematized. The Irish writer Lord Dunsany published a radio play called 'Atmospherics' in 1937, in which a train passenger finds himself alone in a compartment with an escaped lunatic who believes he is receiving radio transmissions telling him to murder his travelling companion. He buys time by persuading him to wait until the order has been repeated three times: '"Are you quite sure that wireless brain of yours is in good order this morning? What I mean is, are you sure it's not atmospherics? You know the very best sets do sometimes."'[68] Eventually, the passenger escapes by pretending to be the lunatic himself so that he is taken off the train. We assume that the crossed wires will be sorted out, though the sketch ends with the transmission being swamped in intradiegetic noise: '*The rest of the conversation is drowned in the triumphant exultation of a train leaving a platform*'.[69] Another Dunsany radio play, 'The Seventh Symphony', takes us into the dream of a delirious composer who has a conversation with Beethoven, Shelley and Keats in which they urge him to leave earth behind. He has arranged with his landlady to be woken by a scheduled broadcast of Beethoven's Seventh Symphony, but is plucked back from eternity by a mistuning, which plays a raucous jazz programme – a kind of human static – which breaks into and wakes him from his ethereal ecstasy:

> I can barely hear Earth now. How rich are the colours of sleep. That's not the Seventh Symphony! What's that? I'm waking! Oh, I'm waking. All the colours are fading. Masters, I don't want to go! They're waking me up with the world's noises; with all the clatter of Earth.[70]

Radio places the sublime and the debased adjacent to, or just the click of a switch away from, each other. The atmospherics now break in, not from outside the transmission, but from within it, as part of the spectrum of radio frequencies.

When the reviewing of broadcast music became an established part of musical journalism, complaints about the degradation of music by the hit-and-miss conditions of radio reception were common. 'We have all', complained one reviewer, 'at times had to lay aside the head-phones in disgust or in despair, because of the hoots and shrieks, groanings, hisses, and gurglings that proceeded from them'.[71] During this period 'the word 'atmospherics' was sometimes carried across from the material context into the content of the music being transmitted. In his inaugural column for the *Musical Times* in 1929, the radio reviewer 'Auribus' reported that a broadcast of chamber music he had recently heard 'was comparatively free from atmospherics, although the principal composer was Schönberg and the others were two of his pupils'.[72] He maintained the blurring of medium and musical form to the end of his review:

> We seemed to be spending our time on the verge of suicide or the end of the world, listening in suspense while throttled words were wrung from a sibylline voice to the accompaniment of apparently idle successions of notes and chords. I give up Schönberg, weakly, if you like, to people who can more easily get his wave-length.[73]

Siegfried Sassoon wrote similarly, in 'Le Sacré du Printemps', a poem about a performance of Stravinsky's *Rite of Spring*, of 'vibro-atmospheric copulations/With mezzo-forte mysteries of noise'.[74]

Etherphonics

From the 1930s onwards there were growing efforts to control atmospherics, not by expelling, but by incorporating them. It was during the 1930s that film and radio technicians began the practice of capturing and manipulating 'room-tone' and environmental backgrounds, using what became known as 'atmosphere microphones' in order to gather these bottled atmospherics. A number of composers literally did begin to try to harness and orchestrate the sounds of the atmosphere, and to develop methods and musical languages in which atmosphere was integrated into the foreground form of the music. One of the most notable of these was Edgard Varèse, the title of whose *Ionisation* of 1929–31, scored for percussion instruments alone, hinted at a radio context. In his attempts to widen the scope both of music production and forms of listening, Varèse incorporated music for an instrument that dramatizes the entire spectrum of attitudes towards the idea of auditory interference and atmospherics: the theremin.

Lev, later Leon Theremin, was a Russian radio scientist who was working on devices that would automatically sense human bodies when he discovered that, by introducing his hand into a tuned circuit involving a gas, he could induce a change in the capacity of the circuit, which altered the pitch of the tone that the circuit delivered. From this was born the idea for an instrument that could be played by hand movements in the air alone. Soon he had added a second circuit, a horizontal loop to be manipulated by the left hand, which controlled volume. By 1920 he had completed the first working version of an instrument he called the 'etherphone'. He quickly became a celebrity in his native Russia, at that period still enthusiastically encouraging technological invention. In 1927 Theremin set out to demonstrate his instrument in a series of concerts and performances in Europe and the UK.

The instrument caused rapture and suspicion in equal measure. Some saw in the new instrument an actualization of the desire to escape the fixed pitches and intervals bequeathed by the Western musical tradition. With an instrument like the theremin, as it was now increasingly known, it was possible to play between established pitches and colours. It was an instrument of in-betweenness, the musical equivalent of tuning between stations, in a kind of free, as yet unpopulated and uncharted radiomusical space. In a sense it was pure atmospherics, promising a world where there would be no necessity for instruments at all. Later in his life, Theremin would experiment with instruments that could be played simply by movements of the eyes, or even by thoughts alone. Such fantasies had already been encouraged by early experiments in wireless telegraphy, which suggested that, if electrical impulses could travel through thin air without the need for any intervening medium, perhaps the body could also be removed from the circuit. Henry Highton, who experimented with his brother Edward on wireless telegraphy in the 1860s, reported in his paper on 'Telegraphy Without Insulation', read before the Society of Arts on 1 May 1872, on an apparatus that employed a thermophile, which created current from variable heat: 'You may judge of its delicacy when I show you that the warmth of the hand, or even a look, by means of the warmth of the face turned towards a thermophile, can transmit an appreciable signal through a resistance equal to that of the Atlantic cable.'[75]

And yet, the music that the theremin produced was a product not of removing the body from the circuit, but of introducing the body into it. In a sense, it was all interference. The report of Theremin's Paris performance that appeared in *The Times* emphasized this, explaining the workings of the instrument by referring to the experience of 'users of valve sets [who] are familiar with the phenomenon of "howling" which occurs as the result of electrical oscillations under certain conditions'.[76] Early audiences were frequently reminded that the sensitivity of the instrument made it apt to produce alarmingly grotesque noises. Reporting on Theremin's lecture-performance in the Albert Hall in December 1927, the *Birmingham Post* was struck by the 'examples of mere noise . . . highly suggestive of the range of tones obtainable from the "taming" of the wireless "howl"' (quoted, *Theremin*, 66). On the one hand, the theremin was capable of tones of an ethereal purity that was hard to achieve with any other instrument; on the other hand, the 'cracklings and buzzings' and the 'stray bleats and wheezes' (quoted, *Theremin*, 60, 251) were only a twitch of the fingers away. *Time* magazine would later describe Theremin as 'the Russian who makes music out of radio static', while Samuel Hoffman, one of the later exponents of the theremin, himself described his technique as 'controlled static' (quoted, *Theremin*, 146, 279). The musical interest in controlling and unleashing feedback that has been a feature especially of rock music, from the 1970s onward, is in a direct line from the theremin.

As Albert Glinsky suggests, the real importance of the theremin may have been the fact that it seemed to make the matter of sound available to be manipulated in an infinite number of ways: 'the raw materials of sound were now exposed and could be molded in every dimension' (*Theremin*, 67). The theremin intimated a world in which music could be anything – and anything could be music. And yet, despite Theremin's own large ambitions for the instrument and the support of pioneers like Varèse, the theremin itself gradually came to be quarantined in a tiny portion of the musical spectrum – its use in films like Hitchcock's *Spellbound* meaning that it was stuck with signifying weirdness, psychic or psychological, and then, from the 1950s onwards, with otherworldliness and extraterrestrial forces. The sleeve-notes to *Music for Heavenly Bodies*, an album of theremin music issued in 1958, promised listeners that it would give them 'the awe-inspiring feeling of asteroids and comets . . . of falling off into the whistling world of infinite space' (quoted, *Theremin*, 290). The theremin arose during a period in which music was opening up to the intrigues and enigmas of unearthly sound; but it played a large part in bringing them down to earth.

Sonification

Radio did not just amplify or enhance, as the microphone or telephone did. Radio belonged to a new, mixed sense, and sense of the mixed, in which oscillations of any kind could be rendered as sound. If one side of this fantasy of interconvertibility was the sonification of matter, the other side was the prospect of the manipulability and transformability of sound itself, which became a kind of ideal, maximally mutable material. According to the *Philadelphia Public Ledger* of 2 March 1928, one of Leon Theremin's more Carrollian competitors claimed that the theremin was an inferior version of a much more powerful instrument of his own invention, which was 'capable not only of producing music . . . but odors and light beams – and conversely, capable of annihilating sound, absorbing it, transforming it into silence' (quoted, *Theremin*, 83).

Digital technology has accelerated the involvement of sound in this kind of intermediality. Everywhere, sound artists are dreaming up ways of using non-sonorous actions, conditions and events to generate sound. In a project entitled *Atmospherics/Weatherworks*, for example, Andrea Polli has developed software that will sonify the data representing storms, cyclones and other dramatic meteorological events.[77] One of the odd things about this project is the idea of replacing or supplementing the natural sounds of hurricanes and cyclones, which one might have thought were more than adequately audible, with sonified data deriving from windspeeds. In part, this is in order to satisfy a desire to hear what can never otherwise be heard (the sound of winds at 50,000 feet). Radio allows one the experience of proximal distance, of being in the presence – the very eye, or ear of the hurricane – of what is physically inaccessible.

Everywhere, it seems, there is the desire to expand the reach of the ear, both in terms of what can be brought to its notice – whether the microscopic munchings of snails, or the howlings of supernovae – and in efforts to integrate into listening the backgrounds and atmospheres that ordinarily remain unnoticed. The contemporary fascination with sonorous immersion and ambience, in atmospheres and soundscapes, which has been so assiduously traced by David Toop, belongs to this ambition to open one's ears to the infra and ultra-sonic atmospheres that surround sound itself, 'to introduce space and air, chance and memory into an otherwise claustrophobic world'.[78]

But the more we enlarge our tolerance of noise, the more we process noise into signal, and therefore make it over into our terms. The sonification of the world is not so much a Cageian tuning into the sounds of things as a modulation of things into our frequency range. For when atmospherics become constructed atmospheres, they are a kind of autistic insulation, as much a way of keeping out as letting in. The rhetoric of the atmospheric enjoins vigilance, exposure, a permission given to the unpredictable; but the more we enlarge our tolerance of noise, our apprehension of the atmospheric, the more it becomes an atmosphere *for us*. If the work of contemporary composers and sound artists seems aimed at finding sonorous correlatives for that which lies beyond or beneath our sensory notice, allowing it to break in upon us as sound, this is accompanied by a denaturing of sound itself. Rather than the spontaneous overflow of meaning and being, sound is just one processing outcome, just one of the many forms into which data can be translated. The press, the presence, the intractable demand on us of sound is being diminished, as the realm of the inhuman has been contracted to the human.

And yet uncertainty, the hum or hiss of background noise, remains, and not just at the edge of the system, but in its midst. At its extreme limit, total information is indistinguishable from total noise. It may be that the alternation between meaning and chaos is constitutive of the kind of intelligence we possess and represent. Aristides, probably writing in the late third or early fourth century CE, offers a musical allegory of creation, in which the body of man is created when the soul descends from the realm of the empyrean, where it is composed of pure geometrical lines and planes, into a condition of materiality. Its materiality is a kind of atmospherics, brought about by the increasing humidity and adulteration of the air through which it falls. As it approaches the airy and humid region of the moon, which makes 'much and vehement whistling because of its natural motion', the soul precipitates a body from wet breath.[79] Here, rather than being the ultimate destination, the resting point or *ne plus ultra* of sound, man, like the early radio operators, is always in the middle of listening. As such, he can come to rest securely neither on the side of information nor of noise, neither of signals nor atmospheric, but is a transformer, who repeatedly recreates the difference between the two, and is himself the precipitate of crossed lines, of interference, of atmospherics.

eight

A Grave in the Air

This book has encountered many forms of the paradox that the air has for so long been a vehicle for the idea of the immaterial, of that which exists without material form, even as our very notions of what matter is and does are thought increasingly by reference to the air. The air is as good a representation as we have of the way in which we are formed and borne down upon by what isn't there. The fact that air forms our relation to the absent, and is the palpable form of absence itself, means that it is associated not just with divinity and the demonic, but also with death. In this chapter, I consider the ways in which death and air have become intermingled, in what may be called the mortification of air.

Dying into Air

Let us begin by observing briefly two contrasting understandings of the relation between death and air. The first is to be found in Sophocles's *Antigone*. The beginning of the play finds Antigone determined to disobey the edict of her uncle Creon, that her brother Polyneices should lie unburied on the ground, unhonoured by any funeral rite, for the birds and dogs to consume. A little later, we have the report of a guard that somebody has disobeyed him and sprinkled the body with 'thirsty dust'. At Creon's command, they re-expose the corpse and retire to a nearby slope, upwind, to escape its putrid stench and keep watch for the renegade sexton. At this moment, there is a dust-storm:

Suddenly, a tornado struck. It raised dust
All over the plain, grief to high heaven.
It thrashed the low-lying woods with terror
And filled the whole wide sky (*en d' emestôthê megas aithêr*).
We shut our eyes
And held out against this plague from the gods.[1]

The failure to bury the corpse, leaving it open to the air, results in a pollution of that air, which is then filled or choked with dust. This idea recurs later in the testimony of Teiresias, who, hearing a savage screaming of birds, tries to make a sacrifice to the gods. Where earlier there had been a contraction of expanse, now there is a failure of height:

I tried burnt sacrifice.
The altar had been blazing high, but not one spark
Caught fire in my offerings. The embers went out.
Juice was oozing and dripping from thighbones,
Spitting and sputtering in clouds of smoke.
Bladders were bursting open, spraying bile into the sky (*kai metarsioi cholai diespeironto*);
Wrappings of fat fell away from soggy bones.[2]

The pattern is clear: what should go up in holy smoke (the fat that was sacred to the gods) fails and slides heavily downwards: instead, the gall and the dust rise into the air, making it impossible to breathe, and impossible to make out a meaning, though it is augury enough of the vitiation of the air. Ooze replaces blaze, the *visqueux* invades the vertical. Teiresias explains that the pollution extends far beyond the unburied corpse, because the birds that have hacked away at it have sullied the hearths of the population. If this were a structuralist reading, the bird might be expected to figure as the principal mediator between the semiotic zones of life and death, earth and air, purity and corruption.

The second concerns a more recent funerary rite, that conducted by Philip Pullman for his stepfather in May 2002. Here is his explanation:

> 'We couldn't decide,' says Pullman. 'What should we do with the ashes? Bury him at sea, scatter him over the hills? No particular reason to do either . . .' 'I thought,' Pullman continues, his tones level, telling the tale of the ashes, 'wouldn't it be a good idea to send him up in a rocket, in a firework? And the others all thought, yeah, what a good idea. So my sister – who knows absolutely everyone who's anyone – found a firework-maker in Edinburgh, and said, "Can you help?" and he said "Yes."
>
> For the quantity of ashes, the firework-maker made up a consignment of 40 rockets, dispensing ashes by spoon into each firework. 'And you know,' says Pullman, 'it was great. We said a few words and then lit the rockets, and up they went, and it was the most wonderful display, and the sky was full of dad, full of stars.'[3]

Pullman's account makes this sound like a very classical affair, for classical myth is full of stories of lovers and heroes who end up immortalized as constellations. But in reality, it is the opposite. To become a star or a constellation is to be fixed, to undergo a kind of stellar petrification, since, for the ancient world, the stars were eternal and immutable. Pullman's stepfather has been scattered, atomized, dispersed, in a kind of apotheosis into mutability. In *Antigone*, we have the horror of the failure to keep apart the dead and the air; in the ethereal interment of Pullman's stepfather we have an ecstatic, triumphant coalescence of the air and the body of the dead.

The desire for air burial is steadily, so to speak, gaining ground. Space Services Inc. of Houston, Texas, launched the ashes of Timothy Leary and Gene Roddenberry, the begetter of *Star Trek*, into orbit in 1997, and they have since been followed by many others who have made arrangements to have their ashes orbit the earth permanently, or be a second time consumed by the heat of re-entry in the atmosphere.[4] Following his wishes, the ashes of Hunter S. Thompson were mixed with fireworks, packed into 34 mortar tubes, and, on 20 August 2005, blasted out across his estate in Woody Creek, Colorado, to the sound of Norman Greenbaum's 'Spirit in the Sky'.[5]

Cremation's in the Air

The rapture that attends the thought of dying into air perhaps begins, or comes to a focus, with the conspicuous valuation of the air that is such a feature of Romantic thought and writing. Reflecting in the final lines of his 'Ode to a Nightingale' that 'Now, more than ever, seems it rich to die', Keats petitions: 'take into the air my quiet breath'.[6] One of the ways in which the Romantic trope of death into air might be thought to be embodied is in the story of the cremation of Shelley, as it is conveyed by Edward Trelawny, an ex-midshipman who was taken up into the Shelley-Byron circle in the 1820s. After his drowning and the discovery of his mutilated body washed up near Via Reggia, Shelley's remains were temporarily buried. It was planned that he should be buried permanently near his friend Keats in Rome, but the bodies of Shelley and his crewmate Edward Williams were in too advanced a state of decomposition, besides which the strict Italian quarantine laws would have prevented their importation or transit across Italian territory. So it was decided to cremate their remains. Trelawny had a furnace made of sheet iron and iron bars supported on a stand. The two bodies were cremated separately, *in situ,* in the two places where they had been washed up. Trelawny goes into some detail about the practical difficulties he encountered. Despite his efforts to fudge up a Hellenic ceremony, the whole enterprise seems to have come somewhat unstuck:

> After the fire was well kindled we repeated the ceremony of the previous day; and more wine was poured over Shelley's dead body than he had consumed during his life. This with the oil and salt made the yellow flames glisten and quiver. The heat from the sun and the fire was so intense that the atmosphere was tremulous and wavy. The corpse fell open and the heart was laid bare. The frontal bone of the skull, where it had been struck with the mattock, fell off; and, as the back of the head rested on the red-hot bottom bars of the furnace, the brains literally seethed, bubbled, and boiled as in a cauldron, for a very long time.[7]

Only Trelawny had the stomach to keep vigil by this inelegant bardic barbecue. Leigh Hunt cowered in his carriage with the windows up, and Byron went for a swim. Nevertheless, Shelley's bungled immolation would come to be identified as the first cremation of the modern era. It is true that Thomas Browne recommended a return to cremation in his *Hydriotaphia* (1658), and a certain Mrs Pratt was cremated after her death and according to her own instructions at Tyburn in 1769. But during the eighteenth century cremation remained tainted by its associations with judicial burnings, which were abolished only in 1790. The burning of bodies after execution was sometimes decreed, often in the case of women convicted of forgery (the last woman to be publically burned in this manner in England was Phoebe Harris in June 1786.)[8]

Although Shelley's Gothic bonfire seems closer to Teiresias's defective sacrifice than Pullman *père*'s skyey apotheosis, it would be easy to see cremation as a more rational form of the Romantic desire to die into the sky, to attain the condition of refinement into holy smoke. One reformer would write that:

> Never could the solemn and touching words 'ashes to ashes, dust to dust,' be more appropriately uttered than over a body about to be consigned to the furnace; while, with a view to metaphor, the dissipation of almost the whole body in the atmosphere in the ethereal form of gaseous matter is far more suggestive as a type of another and a brighter life, than the consignment of the body to the abhorred prison of the tomb.[9]

The pyrotechnical aspirations of Philip Pullman and Hunter S. Thompson are also anticipated in a poem called 'A Request', which appeared in August 1886 in an American magazine called *The Modern Crematist*, in which the sight of a Fourth of July fireworks display provokes mortal yearnings in a pair of lovers (mixed up slightly uncomfortably with other kinds of desire, in a manner somewhat reminiscent of Leopold Bloom's pyrotechnically mediated bliss on Sandymount Strand):

Fizz-Bang! The rockets whizzed aloft,
And on they sped with ease.
'I love,' she whispered low to Jack,
'Such fiery works as these.'

Oh, see that stick in swift descent,
It's fallen on the lawn.
'I dote on pyro-technics, love;
Cremate me when I'm gone.'[10]

In 1890 the Sydney *Evening News* printed a poem that sported, somewhat more sarcastically, with the aerial associations of cremation:

Let's build a crematorium;
Cremation's in the air;
'Twould only cost a trifling sum
To send our corpses there . . .

'Twere pleasant to be wafted hence
As superheated gas,
And so attain the 'burn' from which
No traveller can pass.[11]

Laurence Binyon's poem 'Shelley's Pyre', which is cast into the form of a drama of the elements, puts into the mouth of the 'Spirit of Air' a speech that makes clear the Romantic dream of fiery dematerialization:

The white body is changing: it has taken the swift shape
Of fire, and the fire passes, dazzling the noon,
Shedding all but swiftness and the ecstasy of flight,
Of the light into light.[12]

However, the relations between cremation and air are more complex than these formulations might suggest. The two principal forms of disposal of the dead, burial and cremation, are equivalent in that they both aim to protect living humans from the spectacle of the decaying body: burial by putting it out of sight, and cremation by accelerating the process of decay. In other respects, burial and cremation represent sharply contrasting attitudes to the body. Burial, which established itself almost universally across Europe and America only with the growth of Christianity, and even then only in its second millennium, is seen as a conserving of the body, a sort of earthy incubation against the day when it will rise again. The Pauline metaphor of the dead as the seed from which God will take his harvest at the resurrection rhymes with pagan vegetation and fertility myths, as embodied in a song like 'John Barleycorn', which assign to the buried body the power to regenerate itself in the earth as plants do. Burial provides the illusion of storage, of arresting the effects of time, as though it were equivalent to embalming or mummification, as, of course, under certain conditions of soil and climate, it can be.

Challenges to the practice of burial began only in the nineteenth century, with the rise of the cremation movement. The most important formative pressure was the growing urban population. In the first three decades of the nineteenth century the British population increased by more than 50 per cent, with most of the increase in towns. The resulting overcrowding affected not just the living but also the dead. Intense concern grew during the 1840s about the dangers of disease and contamination posed by the many central London graveyards, and there was pressure to replace 'intramural interment' with burials in more outlying areas where the living need not be in such uncomfortable proximity with the dead.

The argument reached its height during the sanitary 1840s. As part of his energetic enquiry into public sanitation, Edwin Chadwick published *A Special Enquiry into the Practice of Interment in Towns* in 1843. This was followed by a public commission which produced in 1850 a *Report on a General Scheme for Extramural Sepulture*. The concern of the sanitary reformers who successfully campaigned against intramural interment during the 1840s was with the ways in which the dead could transmit disease to the living, and in particular through the production of noxious or poisonous gases. There was concern about the pollution of wells and watercourses too, but this was much less intense than the concern about 'cadaveric emanations',[13] perhaps because it seemed so much harder to keep the delinquent element of air under wraps than water, the flow of which could be controlled and channelled by drainage and the judicious adjustment of gradients. This period was also the climax of the miasma theory, which suggested that disease was principally carried and transmitted by bad air.

These concerns run through Dickens's *Bleak House* of a few years later, in which the circulation of vitiated vapours seems to shadow the movements of the plot and characters. The novel is governed by an air that is both asphyxiatingly effete and yet also dangerously unstable, liable to explosions which, far from clearing the air (an expression that derives from the practice of firing guns in order to drive away pestilence), disseminate its poisons. Nevertheless, concerns about the pollution of water and air by decomposing corpses had begun to be addressed twenty years before, with the rustication of the urban dead in outlying areas away from the most densely populated areas of London. The 1830s saw the opening of the seven major London cemeteries: Kensal Green, West Norwood, Abney Park, Nunhead, Tower Hamlets, Brompton and Highgate. Apart from their location, the other feature of these cemeteries was the fact that they were run by private companies, introducing a link between corpses and commerce that would recur throughout the century.

However, by the 1870s, the explosive growth of the suburbs began to suggest to many that these extramural retreats were soon likely to find themselves once again cheek by jowl with areas of dense population. A more radical solution was announced in an 1874 essay which inaugurated the modern cremation movement and remained its credo for decades. The essay was by Sir Henry Thompson, the physician to Queen Victoria, and appeared in January 1874 in the *Contemporary Review*.

Thompson begins by describing the normal, natural conditions that ensure the recycling of animal remains into organic life. He views this process as taking place essentially, not in the churnings and mulchings of the earth, but in the atmosphere, in which 'those elements which assume the gaseous form mingle with the atmosphere, and are taken up from it without delay by the ever open mouths of vegetable life. By a thousand pores in every leaf the carbonic oxide which renders the atmosphere unfit for animal life is absorbed' ('Treatment of the Body', 319). The atmosphere was the medium of interchange between animal and vegetable, a vast system of reverse respiration, in which the air is 'polluted by every animal whose breath is poison to every other animal being every instant purified by plants, which, taking out the deadly carbonic acid and assimilating carbon, restore to the air its oxygen, first necessary of animal existence' ('Treatment of the Body', 321). Thompson offers an egalitarian vision of the universal convertibility of organic and inorganic forms:

> Our mahogany of to-day has been many negroes in its turn, and before the African existed was integral portions of many a generation of extinct species. And when the table, which has borne so well some twenty thousand dinners, shall be broken up from pure debility and consigned to the fire; thence it will issue into the atmosphere once more as carbonic acid, again to be devoured by the nearest troop of hungry vegetables, green peas or cabbages in a London market garden – say, to be daintily served on the table which now stands in that other table's place, and where they will speedily go to the making of 'Lords of the Creation.' And so on, again and again, as long as the world lasts. ('Treatment of the Body', 320–21)

However, Thompson also inherits another strain of argument from the sanitary reformers of thirty years earlier, for whom, as we have seen, it was the gaseous emanations from the decomposing body that represented the greatest danger. This sharply contradicts this confidence in the power of the air to purge and transform. In a supplementary article, Thompson quotes the words of Lyon Playfair from the 1850 *Report on a General Scheme for Extramural Sepulture*:

> I have examined various churchyards and burial-grounds for the purpose of ascertaining whether the layer of earth above the bodies is sufficient to absorb the putrid gases evolved. The slightest inspection shows that they are not thoroughly absorbed by the soil lying over the bodies. I know several churchyards from which most foetid

> smells are evolved; and gases with similar odour are emitted from the sides of sewers passing in the vicinity of churchyards, although they may be thirty feet from them. ('Cremation', 557)

With his customary thoroughness, Thompson estimated that the 52,000 annual interments in London in 1849 produced 784,122 cubic m of gas, which either passed into the air or the water ('Cremation', 557). Like many others, Thompson was fond of telling stories about the mortal consequences of coming into contact with putrefied air – telling us, for example, that two grave-diggers expired after entering a tomb in St Botolph's, Aldgate ('Cremation', 557). So if, in one sense, the air can be relied on to dissipate the menace of the dead, in another sense, the air is the very theatre of their vaporous mischief. There were other dangers, apart from poisoning, which are also hinted at in the combustibility of Dickens's *Bleak House*:

> The ordinary system of Burial is to seal up the dead in leaden coffins, which are rendered as impervious to the air as possible, but from which the gases manage slowly to escape, or if hindered at first break forth at last with incredible violence, sufficient sometimes to burst open the shell with a noise that has been compared to the report of a cannon.[14]

So two contrasting impulses and attitudes towards the air run through Thompson's writing and the writing of those who follow him. On the one hand, there is the desire to commit mortal remains to the air, the power of which to dissipate, transform and re-circulate is implicitly trusted. This, of course, draws heavily upon the strong Victorian faith in the purifying and prophylactic powers of ventilation and the open air. The coffin (and in the years in which the fear of premature burial thrived, existence in the coffin was never more vividly imagined) was regarded as the very paradigm of air poisoned by being withdrawn from circulation. On the other hand, a more traditional apprehensiveness remained that, if brought into contact with the air, the dead would spread their pollution far and wide.

Thus, even in cremation, measures were needed to ensure the sequestering of the corpse. One of the problems of cremation was that the process of burning actually threatened to produce more noxious vapour in a shorter amount of time than a decomposing corpse. Thus, 'the atmosphere will be rendered injurious to the living through the addition of smoke and gases in enormous quantity'.[15] Thompson was naturally keen to minimize this threat. An early version of the furnace that he favoured employed a secondary furnace to ensure that all the gases produced in the first few minutes of burning, the time of their most vigorous production, would be consumed rather than conducted out of the chimney into the air ('Cremation', 563). An improved version of the furnace, manufactured by William Siemens, employed a filtration system, which Thompson carefully describes:

> In this case, the gases given off from the body so abundantly at first, pass through a highly heated chamber among thousands of interstices made by intersecting fire-bricks, laid throughout the entire chamber, lattice-fashion, in order to minutely divide and delay the current, and expose it to an immense area of heated surface. By this means they were rapidly oxidised, and not a particle of smoke issued by the chimney: no second furnace, therefore, is necessary by this matter, to consume any noxious matters, since none escape. ('Cremation', 563–4)

Much effort was also expended to ensure that no foreign matter was mingled with the ashes of the dead, extending even to the use of magnets to remove any flakes of metal from the interior of the furnace that might have contaminated the corpse during the process of burning. Thompson assured his readers that

> The inner surface of the cylinder is smooth, almost polished, and no solid matter but that of the body is introduced into it. The product, therefore, can be nothing more than the ashes of the body. No foreign dust can be introduced, no coal or other solid combustible being near it: nothing but heated hydrocarbon in a gaseous form and heated air. ('Cremation', 563)

The limits to the embrace of the dissipated body are also suggested by the debates about what one was to do with the ashes of the deceased. Scattering, on sea, land, or even in the air, was encouraged by early cremationists, for whom ploughing the elements of the body back into natural process was the most important consideration. But, as cremation slowly became more popular, it became more and more common to keep the ashes of the deceased together, in urns, or even to bury them in graves. The economic motive was to the fore here, as funeral directors sought to secure the revenue that was increasingly being lost to them with the turn away from costly coffins and headstones. As Stephen Prothero shows, although in theory it was possible for relatives simply to carry away ashes and dispose of them privately, funeral directors adopted various measures to discourage this, including refusing to subject the ashes supplied in this case to the process of grinding, which is necessary to reduce the odd bits of knuckle and bone to the seemly, airy talc of the popular imagination.[16]

A Question of Time

Cremation was promoted as being safe, hygienic, efficient and rational. More than anything else, it was 'a question of time' ('Cremation', 559). Thompson promised that 'by burning, we arrive in one hour, without offence or danger, at the very stage of harmless result which burying requires years to produce' ('Cremation', 560). It was the defeat of disease by speed. Seen in these terms, cremation was a distinctly modern way of death, whose time had distinctly come. The most important form of the claim that cremation was to be regarded as modern was the strong emphasis on the economics of matter. Almost the worst effect of burial, as far as Thompson was concerned, was that it removed valuable materials from circulation for an indefinite amount of time. Thompson emphasized that burial thwarted the economic purposes of nature, the capital of which 'is intended to bear good interest and to yield quick return' ('Treatment of the Body', 85). One of his followers enlarged upon this theme in 1886:

> What would we think of a man who with a view to becoming rich kept burying in the earth greater and greater sums of money, and taking out in place of them proportionately smaller sums which his father and grandfather had buried there before him? We should laugh at such a person, I suppose, and suggest to him to lay out his surplus income to interest. But the conduct we ridicule is our own. We forget that nature allows not only interest but compound interest for invested capital. We reject her proffered usury, and deliberately hide a precious talent in the earth.[17]

Many were horrified by this utilitarian, even economistic language, among them Philip Holland, who wrote a reply to Thompson in the *Contemporary Review* for February 1874, in which he protested at the suggestion that 'we should use our fathers' ashes as turnip-dressing' and jeeringly parodied Thompson's calculation that the import of foreign bones to make good the unavailability of the native variety cost half a million pounds per annum ('Treatment of the Body', 87):

> Why not, as we easily might, dry and reduce to powder the flesh as well as the bones of our relatives and friends, to be used as a substitute for guano – thereby saving the whole amount of £700,000 a year, which, divided amongst the thirty million inhabitants of the British Isles, would amount to the magnificent sum of sixpence a head every year, obtained at the trifling cost of outraging family affection, and desecrating what most of us regard with tender reverence?[18]

Cremationists emphasized other aspects of the modernity of cremation. It was not only economically prudent, it was also hygienic. Campaigners like Thompson stressed the fact that in cremation 'No scents or balsams are needed, as on Greek or Roman piles, to overcome the noxious effluvia of a corpse burned in open air' ('Treatment of the Body', 325). Compared with the barbaric outdoor festivals of crackling bones and spattering suet, cremation aimed at being a thoroughly intramural affair. No doubt the campaign against the practice of suttee that, a century before, Samuel Johnson had sardonically described as the 'custom of voluntary cremation . . . not yet lost among the ladies of India', which had been abolished as long ago as 1829, was still an active ingredient in the revulsion against burning *en plein air*. Cremationists also attempted to keep at bay some of the infernal associations of flame and smoke by emphasizing that the corpse was untouched not only by human hand, but also by any flame, being reduced to ash in a process of baking by incandescent air, the effect of which was to produce what was sometimes described as a halo of ethereal radiance, or a 'cheering glow'.[19]

However, others emphasized the ancient lineage of cremation. Protestants sometimes regarded it as a way of burning off the last clinging wisps of Catholic superstition about the indwelling of the spirit in the body and the resurrection of the flesh. (Browning's Bishop of St Praxed's, looking forward to a posthumous existence snuffing up 'Good strong thick stupefying incense-smoke', is part of this horror of the bad air associated with Catholicism.[20]) Religious radicals and dissidents went even further and blamed the unwholesome practice of burial on Christianity as such, influenced by Judaism. In fact the Old Testament offers no clinching preference for either burial or burning. It is true that burning is threatened as a punishment for certain sexual crimes – Genesis 38.24 and Leviticus 20 – but there is no suggestion that the burning of the bodies of Saul and his son by the inhabitants of Jabesh-Gilead was intended as anything other than honouring them. Elizabeth Bloch-Smith suggests that the population of Jabesh-Gilead may have been Phoenicians, who introduced the practice of cremation into the region.[21] For these, cremation was not so much a rational modern invention as a return to more ancient, pre-Christian traditions, whether these were Nordic or Hellenic (Trelawny would have had some Greek verses intoned over the burning body of Shelley, were it not for the fact that the only one of the assembled company who knew any Greek was no longer in any condition for oratory).

One of the most important steps in establishing the legal basis of cremation arose from a private cremation that was undertaken by a rather wild character called Dr William Price of Pontypridd, a physician, Chartist, vegetarian, nudist and Druid. When, in 1884, Iesu Grist, the illegitimate child he had fathered at the age of 83, died at the age of five months, he announced that he would cremate his remains on a pyre, in accordance, as he believed, with ancient Celtic tradition. A large, and largely disapproving crowd gathered, and it seems that the body was snatched from the pyre and Price himself arrested and charged with illegal disposal of a body. He appeared before Justice Stephen at Cardiff Assizes, who declared that, provided it constituted no nuisance, cremation could not be regarded as illegal.[22] This broke the stalemate that had arisen between the Cremation Society, headed by Henry Thompson, and the home secretary and led the way for the legalisation of cremation in the Cremation Act of 1902. The mock obituary for English cricket which appeared in the *Sporting Times*, following the home country's defeat of England at the Oval in August 1882, in which it was solemnly announced that 'The body will be cremated and the ashes taken to Australia', thus occurred at a time when cremation was an extremely hot topic.[23]

Cremation was embraced by some of the new forms of spirituality and unorthodox religion that appeared in the late nineteenth century, especially those, like Theosophy, which drew heavily on Eastern religions, such as Buddhism, Hinduism and Sikhism, in which the burning of bodies features. Interest grew in more unusual practices, too, like the Tibetan practice of sky burial, and the somewhat more ceremonious version of this practised by the small community of Zoroastrians who found their way to India and, known now as Parsees, clustered around Bombay. Eschewing the rather grisly practice of carving up the corpse into bite-sized portions that characterizes Tibetan sky burial, Parsees leave their corpses to the offices of the vulture in specially constructed dakhmas or 'Towers of Silence'. In 1886 the traveller Mrs Howard Vincent explicitly connected this practice with funerary reform in the West:

> In these latter days when over-crowded cemeteries and the levelling of graveyards in the midst of our metropolis have called forth the cry of 'ashes to ashes, dust to dust,' by some new means, and some means quicker than the old; when even cremation has come within the bounds of possibility, surely the Parsee mode of burial will commend itself to many foreseeing minds. True that we do not like to think of the vultures hovering around the funeral procession

> for the last few miles, nor of others awaiting it, perched on, and greedily gazing down into, the tower.; but is it so much worse than 'the millions of insects of the ground' of our burial, of which the Parsee speaks with such horror?[24]

Mrs Vincent goes even further in reassuring her readers as to the dignity of the Parsees, who are 'most enlightened and civilized, and not to be named with the Hindus. They are European in comparison. And, without doubt, it is in great measure owing to their true and moral religion, of which the rite of burial – the Tower of Silence – is the most beautiful treasure'.[25] These remarks belong to the high point of the nineteenth-century theory of the origins of civilization among the so-called Aryan peoples of the Indic basin (the Zoroastrians, of whom the Parsees are the Indian representatives, arose in Iran, the name of which means something like 'the Aryan land'). Though Zoroastrianism forbids cremation, on the grounds that it defiles the sacred element of fire, the Parsees are presented as the keepers of an ancient flame with regard to the disposal of the dead.

An Air That Kills

The Romantic desire for assimilation into the air is a desire to assume illimitability. By contrast, the cremation movement was part of an opposite process, of occupying, capitalizing on and habituating the air. By the end of the twentieth century this would lead Salman Rushdie to describe the air as 'that soft, imperceptible field which had been made possible by the century and which, thereafter, made the century possible, becoming one of its defining locations, the place of movement and of war, the planet-shrinker and power-vacuum, most insecure and transitory of zones, illusory, discontinuous, metamorphic'.[26] The most important form of the occupation of the air is the development of air power for the purposes of war. Not only does dying into the air become more and more thinkable and even attractive, the prospect of death that comes from the air is ever more imminent and actual for ever larger populations. Archers, fusiliers and bombardiers had made ever greater incursions into the air (the shells projected by 'Big Bertha' during the First World War rose 25 miles before coming to earth). But the middle of the twentieth century saw the first systematic use of what we have come to call 'air-power', along with poison gas, chemical and biological warfare, radiation, chemical warfare. But these are all things that use the medium of air. Of all these, none had more impact than the development of poison gas.

The idea of using poisonous or noisome gases, vapours and smokes to defeat or incapacitate one's enemy is recorded from classical times. For centuries, the favoured substances were pitch and sulphur, or brimstone. Thucydides recorded that the Spartans used arsenic smoke during the Peloponnesian War in the fifth century BE. Leonardo made plans for smoke weapons formed of sulphur and arsenic dust. The first stink bombs or stench bombs were indeed weapons employed for military purposes. One Fioravanti of Bonomia made stench bombs from an oil brewed from a mixture of turpentine, sulphur, asafoetida, human faeces and blood.[27]

This lineage seems appropriate, for there is something archaic in the very nature of poison gas. Bad or lethal air has usually been thought of as emanating from nature, rather than as the result of human device. The word 'influenza' preserves the belief in the malign influence of the stars, transmitted in the form of mephitic fluid or vapour. The fascinators or bewitchers of the medieval and early modern imagination were believed to have the power to blast and wither crops and cattle with their breath, often working in conjunction with the power of the evil eye. The basilisk, which could both immobilize its victims with its eye and destroy them with its mephitic breath, is the mythical embodiment of this belief. Hell, or the underworld, is regarded in many cultural traditions as a stinking or smoky place.

If there is something archaic about poison gas, it is also true that it is a preeminently modern weapon, because of its association with technological development and industrial production. Siegfried Sassoon was one of those who experienced the First World War as the passing away of a Romantic ideal of bravery amid the processes of mechanized death, writing of how, by the winter of 1916–17, 'the war had become undisguisedly mechanical and inhuman'.[28] He may have had in mind the massification of men and the use of aircraft and tanks, though perhaps gas played a considerable part in this mechanization too. Cultures who have developed techniques for smelting and other industrial processes requiring combustion are all familiar with the noxious or toxic by-products of these processes. Mining and cave exploration brought acquaintance with the two ways in which air can be lethal: explosion and asphyxiation. The lesson of the First World War was that only countries with advanced chemical industries were able to deploy gas in a systematic way in combat. There is a particularly long and close association between the dyeing industry and the production of gas. Partly because of the use they made, well into the seventeenth century, of human products like earwax and urine, dyeworks were renowned as extremely smelly places and during the medieval period were often, like tanneries, banished to the outskirts of towns. The principal use for chlorine, the gas that was the first to be used during the First World War, was as a bleaching agent. Ironically, bleaching powder would turn out to be the most effective neutralizer of mustard gas. In 1934 F. N. Pickett, who had been involved in clearing the large dumps of German chemical weapons left over after the First World War, and who thought that gas was certain to be employed in any future war, warned that 'Our dye industries, and therefore our poison gas manufacturing facilities, are not among the great industries of the world'.[29] German poison gas production during the First World War was driven by a conglomerate of eight chemical combines in the Ruhr, known as Interessen Gemeinschaft, or IG, who had a world monopoly on production of dyes.[30] IG Farben would be associated with Zyklon B, the most notorious of the gases employed during the Second World War.

Because of its notorious fickleness and the difficulty of deploying it reliably in battlefield situations, or even, as the disastrous outcome of the Moscow theatre siege in October 2002 demonstrated, in enclosed circumstances, the use of gas requires great technical skill and precision.[31] Until recently, the production of chemical weapons such as gas required considerable industrial effort and coordination, with advanced techniques of mass production, storage and distribution, reinforced by well-established scientific infrastructure. It is for this reason that, despite its archaic nature, and its recent associations with small groups or countries attempting to equalize a military disadvantage, gas has tended to be used by countries enjoying industrial superiority over their adversaries.

Although the outcomes of war are determined by such cultural-technological differences, war is also by its nature traditionally supposed to create a kind of mirroring or mutual acknowledgement in the adversary relationship. Gas has a unique reputation for being perfidious, for setting aside the relations of mutual respect and recognition that are supposed to hold even in the most savage and unbridled conflict. An Austrian chemist, Veit Wulff von Senfftenberg, wrote in 1573 about an early example of the stink bomb: 'It is a terrible thing. Christians should not use it against Christians, but it may be used against the Turks and other unbelievers to harm them'.[32] When a French general, Peleesieu, used a cloud of smoke generated by green wood to suffocate a tribe of Kabyis in 1845 in Ouled Ria, he was recalled for what was regarded as an offence against codes of military honour. Lord Dundonald recommended the use of gas against the French in 1811 and again during the siege of Sebastopol in 1845. A committee of enquiry rejected the idea as dishonourable.

Gas appears to have been used against Afghan rebels in the 1920s and certainly was used by Mussolini during his invasion of Ethiopia in 1935. The use of gas shells was condemned by the Hague Conventions of 1899 and 1907. The French had evidently employed gas grenades during 1915, though to little effect. This example of first use may have emboldened the Germans to the politically risky act of retaliation in kind, though their first two experiments in firing gas-filled shells, against the British and Russians, had such insignificant results that neither adversary even realized that gas had been used against them. But then, on 22 April 1915, German troops massed around Langemark in Belgium opened cylinders of chlorine gas that had been carefully placed in their trenches. Wind and weather were kind, and the gas formed a thick greenish cloud which drifted slowly towards the Allied lines. The result was catastrophic for the British and Canadian troops, around 5,000 of whom were killed and many more incapacitated by the choking gas. Doctors had no idea how to treat casualties: one death was diagnosed as due to 'air hunger'.[33] Chlorine attacks the inner lining of the lungs, causing the victims of most serious poisoning eventually to drown in their own exuded fluids. It is the effects of chlorine that lead Wilfred Owen to the apprehension of 'the old lie' in 'Dulce et Decorum Est':

> Gas! GAS! Quick, boys! – an ecstasy of fumbling
> Fitting the clumsy helmets just in time;
> But someone still was yelling out and stumbling,
> And flound'ring like a man in fire or lime . . .
>
> If you could hear, at every jolt, the blood
> Come gargling from the froth-corrupted lungs
> Obscene as cancer . . . [34]

Even though the Germans vehemently denied the first use of gas in the First World War, and insisted that their actions had not been in breach of the Hague Convention of 1907, because this prohibited only the use of gas-filled projectiles, which they had not on this occasion used, their attack was condemned as an unfair and treacherous assault upon an unprepared opponent. It is, however, hard to see why this really differs from the introduction of new forms of artillery or high explosive, or new methods of delivering weapons, such as balloons or aircraft.

To begin with, the Allied troops seemed utterly defenceless against this new weapon. With no other form of protection available, soldiers were advised to breathe through socks soaked in their urine, or even to bury their mouths and noses in earth and use that as a protective filter. Within weeks, however, the British and French had begun to develop protection against gas. If the dominant German chemical industry gave them a conspicuous lead in the production of gas, the long experience of mining in industrial Britain gave them invaluable experience in protecting against dangerous gases. It is the conjunction of explosion and asphyxiation, the two effects of poison gas, that enables Owen to make a striking connection between mining and soldiering in the poem 'Miners', in which the sinister hiss of the coals on his fire make him think that

the coals were murmuring of their mine,
And moans down there
Of boys that slept wry sleep, and men
Writhing for air.[35]

Within months, the British and French not only had respirators to protect against gas, but also had stocks of gas of their own ready to employ. Thereafter, gas was used extensively on both sides. Defence and attack leapfrogged over each other.[36] The first respirators, simple breathing pads secured by tape, were rendered useless by the German introduction, at Ypres on 19 December 1915, of the much more deadly gas phosgene. This gas, which had been discovered by Humphry Davy in 1812, was a mixture of chlorine and carbon monoxide, which was even more noxious than chlorine and dispersed less easily. This led in turn to improved respirators, in the form of gas helmets and hoods. The balance shifted again with the German introduction of mustard gas in mid-1917. Mustard gas (or dichlorodiethyl sulphide) is, like many of the chemical compounds of this period, somewhat misnamed, for it is a liquid at normal temperatures and is dispersed in the form of an aerosol. It is a vesicant or blistering gas, which acts not just on the lungs, but also on any exposed areas of skin, causing intense blistering, especially around the eyes. Sargent's painting *Gassed*, showing blinded men being led in a line away from the battlefield, shows victims of mustard gas. The deployment of mustard gas led to improved vigilance and what was called 'gas discipline'. By the end of the war, the shortage of supplies and the development of new arsenical compounds, such as the American gas Lewisite, in which arsenic was mixed with mustard gas in order to poison the wounds caused by the blistering, had shifted the advantage to the Allies, who were seriously planning the first gas attacks by aircraft on civilians in cities.

The experience of the First World War, and the expectation of gas attack from the air in the early years of the Second World War, meant that Britain gave serious thought to its first use. Sir John Dill, Chief of Imperial General Staff, proposed using mustard gas on an invading German army on 15 June 1940. His idea was vetoed, on the grounds that it would provoke retaliation and mean a loss of moral authority, especially among Americans.[37] A member of Dill's own staff, Major General Henderson, wrote that '[s]uch a departure from our principles and traditions would have the most deplorable effects not only on our own people but even on the fighting services. Some of us would begin to wonder whether it really mattered which side won.'[38] As hostilities intensified, Churchill, who had been involved in planning gas attacks in the First World War, seemed to abandon whatever restraint he had about first use, writing in a minute to his Chiefs of Staff of 6 July 1944:

> It may be several weeks or even months before I shall ask you to drench Germany with poison gas, and if we do it let us do it one hundred per cent. In the meanwhile, I want the matter studied in cold blood by sensible people and not by that particular set of psalm-singing uniformed defeatists which one runs across now here now there.[39]

So poison gas is at once modern and archaic, at once an instrument of war and a betrayal of it. What we call modern warfare is an example of that intensified swirling together of the new and the old that is actually the leading characteristic of what we have perhaps for too long been too content to call modernity. It may be that the particular kind of apostasy represented by poison gas has to do with the way it embodies the fundamental ambivalence in our relations to air, which all of us, at every moment, are at once absorbing in and expelling. Air is both life-giving and noxious. The goodness and the badness of air are intimate opposites, which meet and invert in the intoxications of tobacco, opium and other inhaled drugs. It had become apparent from the early seventeenth century that the dangerous gas produced by the burning of coal and charcoal, and the fermentation of wine, known as carbonic acid gas, or fixed air was produced within the body, as what one late eighteenth-century medical writer called the 'fecal matter of the vascular system'.[40] The same writer reported that the dangers of fixed air could be demonstrated by the fact that '[I]f in the morning a lighted candle is placed under the cloaths of a bed in which a person has lain all night, so great is the accumulation of this gas, that the flame is immediately extinguished.'[41] During the 1930s an anti-war organization called World Peaceways placed an advertisement in various magazines, showing a fiendish-looking boffin, at work upon an annihilating super-gas. The caption threatened that, in the next war, 'Planes will zoom over cities and towns, children will fall down strangling from one breath of air that a second ago had seemed pure and sweet.'[42] Poison gas is air betrayed, not the enemy of life, but life turning on itself.

All this may be thought of as a familiar history, since the technologies of war enter into the fantasies and lived realities of populations, especially perhaps in modern warfare, in which communications and military technology are so closely combined. Poison gas, or the idea of it, is both unthinkable and insidiously familiar. In the interwar years in particular, the idea of gas took on a kind of political and phantasmal reality that has not yet diffused, and serves as a point of reference for more contemporary concerns and debates. One might equally think of the history of gas as an imaginary history, since it is the history of the idea or dream of gas as much as its reality. Just as the old and the new are not easily to be distinguished, so too the idea and the reality of gas form together an unstable yet indissoluble compound. The accounts of Saddam Hussein's gassing of the Kurds and the fear of gas attack and attack by chemical weapons during the first and second Gulf Wars reactivated many of these fears.

L. F. Haber's detailed and almost pathologically sober account of the use of gas in the First World War in his *The Poisonous Cloud* includes a few discussions of the representation of gas by artists and writers. He is bewildered by the fact that so many should have come to see gas as the ultimate weapon and the ultimate atrocity during and especially after the war, since his view is that gas was a peripheral and, in the end, rather a feeble weapon, the effects of which were hugely enlarged by rumour, fantasy and imagination. And yet the psychological component of gas was widely recognized and formed part of the reality of its effect. Gas is a mass weapon in two senses. First of all, it seems to be aimed at mass destruction; but it is also a weapon that is implicated in the epidemic processes of mass media and communications. It is not just the fact, but also the idea of gas, that tends towards a condition of saturation.

One of the virtues of the arsenic-based compounds that began to be introduced in the later years of the First World War was that they included distinct psychotropic effects, inducing depression and panic in addition to the demoralizing fear naturally provoked by the idea of gas. F. N. Pickett wrote that

> The writer has probably been through more gas clouds than any other individual living, and is not ashamed to confess that he has and always has had a dread of poison gas, even though he might know that the gas was not in a dangerous concentration.
>
> This dread is entirely psychological, and a *recognition* of this dread which almost amounts to hysteria is the first step in defeating gas.[43]

This advice comes from a book that is meant to be reassuring, but seems unlikely to have been. Gas seems to have gathered a reputation for having the power not just to injure the body, but also to destroy the moral fabric of the person: indeed, its reputation is part of this power. One anti-war pamphlet from the 1930s claimed that 'The mental distress caused by [gas] poisoning is often sufficiently serious to drive the sufferer actually temporarily insane.'[44] It goes on to evoke

> those tragic human wrecks whose nervous systems have been completely disorganised – and who fill the nervous hospitals. These shadows of what they were are startled, frightened creatures – sleepless and apprehensive – unable to concentrate – often completely losing their memory for a while – at times suicidal, at times unable to walk or move or in any way help themselves.[45]

During the 1960s the psychotropic possibilities of gas were investigated from the other end, with military experiments on hallucinogens like LSD. A report of 16 August 1963 in the *Wall Street Journal* quoted one US government scientist as saying 'Ideally we'd like something we could spray out of a small atomizer that would cause the enemy to come to our lines with his hands behind his back, whistling the

Star-Spangled Banner. I don't think we'll achieve that effect. But we may come close.'[46]

Bruno Latour has suggested that the great founding error in the formation of what he calls the 'Modern Constitution' is the idea that modernity creates an absolute gap between natural existence and human artifice. In fact, modernity, with its host of techniques of technical transformation, does not make the natural world over into the cultural, but rather gives rise to an ever-expanding middle ground of what Latour, borrowing the term from Michel Serres, calls 'quasi-objects'. The principal example that Latour offers for this is in fact gaseous, for it is the vacuum procured by means of Boyle's air-pump.[47] A vacuum is a natural fact, and yet also entirely an artefact of the laboratory. A gas is similarly ambivalent. Many of the gases produced during the First World War and beyond are in fact to some degree naturally occurring and all of them are natural, even as they are clearly human fabrications. As a kind of 'second nature', as artefacts that are nevertheless never entirely controllable, they are quasi-objects *par excellence*.

Gas is never pure, never just one thing. Indeed, few of the weapons described as gases in the history of this particular form of warfare in fact exist as gases at normal temperatures: chlorine and phosgene are the exceptions. Other so-called gases are dispersed in the form of smokes, liquids, powders and aerosols. It is for this reason that the area of technical speciality devoted to this form of warfare quickly became known, as it is still known today, as 'chemical warfare'. And yet, terms like 'gas' and 'gassing' survive, for example, in accounts of Saddam Hussein's gassing of the Kurds. What all these forms of offensive military operation have in common is the fact that they are airborne, carried in, or transmitted in the air.

As such, they belong to a no-man's land. Gas is not so much a weapon, as a form of communication. Weapons are designed to establish in the most unambivalent of terms the difference between subjects and objects. A subject who employs a weapon against another subject makes that subject simply and utterly the object of his attack. The spear is the stereotypical form of the weapon in this. At one end, there is a handle or grip; at the other, the business end, is the sharpened tip.

Ships, shells and bombs continue to approximate to this shape. At one end, in other words, there is the sphere of the subject, at the other, the place of the object, the object you become by being in this place. In fact, this is not quite right, since the concentration of the point at the sharp end is meant to indicate not the location of the aggressor, but his potential omnipresence, along with the reduction to space, to the x that marks the spot, of the enemy. You defend yourself under such conditions, either by repelling the spear or by avoiding it, which is to say, by giving yourself the invulnerability of your adversary. Much of the mythology of gas derives from the improbable piece of meteorological good fortune that allowed the cloud of chlorine released by the Germans on 22 April 1915 to proceed gently downhill and in an easterly direction towards the British troops. Thereafter, the efforts of First World War combatants were directed towards making gas behave as a projectile, not with much success.

A fist, spear, shell or missile cleaves through the air, overcoming its resistance, just as gravity is temporarily set aside. The parabola is the perfect form of the compromise between nature and culture. Camille Paglia sees a principle of transcendence in the male capacity to generate arcs of urine, as compared with the inchoate immanence of the female, forced by the absence of a focusing or sighting organ, to gush where she squatted.[48] She might well appreciate the ballistic finesse of the joke about the drunk who is preparing to relieve himself by a lamppost when he is restrained by a policeman who tells him 'You can't do that here'. 'I'm not going to do it here', replies the drunk. 'I'm going to do it Right, Over, THERE!'

Gas, by contrast, does not have a sharp end, cutting edge or, as we say, 'front line'. Gas does not penetrate, but rather diffuses or infiltrates. The nature of all gases is to expand, uniformly, in all directions. The big problem for engineers of gas attacks was how to balance the need for the gas to diffuse over a wide area, like a slow bomb, while also keeping it sufficiently concentrated for its toxicity not to decline. How does one do battle – in either sense, as attacker or defender – with a cloud rather than a spear? Peter Bamm, a German doctor who published novels and reminiscences under the pseudonym of Kurt Emmerich, wrote that it was impossible to be brave against gas.[49] With gas warfare, the trigonometry that had governed military theory in an unbroken continuity since the days of Roman siege catapults, gave way to a fiendishly complex multi-parameter calculus, in which the simple variants of propulsive force, air resistance, gravity and distance were complicated by the addition of wind direction and speed, air pressure, humidity and temperature, all of which affected the rate of expansion of gases. Neither the meteorology nor the mathematics of the combatants was adequate to the task.

It became apparent both to the Germans and the Allies that if the use of gas clouds were to continue, the advantage would be sure in the end to lie with the Allies, for the ludicrously simple reason that the prevailing winds in northern France do not blow from east to west, but from west to east. This is not to mention the dangers of blowback, which meant that no army could be sure that a sudden change in the direction of the wind would not turn their weapon back on them. Even if the wind behaved properly, there was the problem for the forces following up the gas attack of themselves having to advance through a gas cloud, or take possession of terrain that had been contaminated. Gas was not only dependent on the weather, it also represented the principle of turbulence itself.

I alluded in chapter Four to the distinction that Bruno Latour makes between an intermediary and a mediator. An intermediary is an object that connects two subjects by moving between them. It simply 'transports, transfers, transmits energy'.[50] A mediator is a third thing, which forms an environment between the two subjects it connects. It is 'an original event which creates what it translates as well as the entities between which it plays the mediating role'.[51] It is an atmosphere and not an object. As the metaphor I have just used would imply, gas is this kind of mediator, or, as I called it in chapter Four, a mediate matter. Rather than a missive or a missile, it is what Michel Serres calls a 'milieu'.[52]

Gas became the archetypal weapon of the First World War because it instituted the ghastly economics of exchange between the combatants. Though gas was in no way decisive in the war, it seems clear that it would have been, had the Germans been able to limp on into 1919, at which point the British, assisted by the newly arrived Americans who had poured a great deal of resources into the development of gas weapons, would have had clear gas superiority for the first time. The Germans had already had to replace the rubber of their respirators with treated leather, and would not have been able to maintain supplies of these materials. The design of their respirators – all-in-one snout respirators attached to the face rather than the British design of hood respirators, attached by a tube to a separate box containing the filter mechanism – meant that they would not have been able easily to be adapted to cope with the arsenical compounds, especially Lewisite, which the British and Americans had been developing. Even more banal and most telling of all was the shortage of fabrics, which meant that the Germans would not have been able to replace uniforms and boots contaminated by burning droplets of mustard gas, which had taken over from chlorine and phosgene as the most successful form of chemical warfare during 1916 and 1917.

A curious effect of gas was that, although it turned out to be a fairly simple matter to protect against it, the very measures that should have provided the solidest reassurance in fact intensified the fear. Although most troops at the front did not experience gas bombardment on a regular basis, the necessity of regular drilling in order to ensure adequate gas discipline kept the dread of gas alive. Mass technological warfare requires the maintaining of a difficult affective mixture: there are the traditional daring and conviction required for exposure to shot and shell and the patient, meticulous attention to detail of the operative keeping his machinery in good order. Gas required a terrifyingly continuous attention to one's safety, as focused upon the state of one's equipment, which may in the end have been psychologically draining. The need for inspection is only one form of the way in which gas turns the combatant or other potential victim in on himself, making death or injury one's own responsibility. The respirator on which one's life depends also isolates and disorientates, making communication and the aiming of weapons difficult, as well as cutting one off from what is often the most important of the senses, that of hearing. Saddam Hussein was far from the first to learn the lesson that the very means of defending against gas can become a form of assault. As the destructive limitations of gas became apparent, it started to be used as an instrument of attrition. The fact that gas did not kill as many men as high explosive could be a positive advantage, Twenty per cent of men killed leaves 80 per cent able to fight. Twenty per cent injured will absorb the resources of perhaps a further ten or twenty per cent in evacuating and treating them. On one occasion, a particularly intense British gas bombardment of seven-and-a-half hours duration caused a German withdrawal, not because the gas succeeded in penetrating the German respirators, but simply because the stress and fatigue of wearing them became too much to bear. Wilfred Owen's 'Dulce et Decorum Est' evokes the shared predicament of the gassed man and the masked man:

> But someone still was yelling out and stumbling,
> And flound'ring like a man in fire or lime . . .
> Dim, through the misty panes and thick green light,
> As under a green sea, I saw him drowning.
>
> In all my dreams, before my helpless sight,
> He plunges at me, guttering, choking, drowning.[53]

The striking thing about the poem is the interchange it effects between the victim, drowning in the green cloud of chlorine, and the strange, dreamy, submarine sensation of the one inside the mask. It is safety that seems suffocating.

This strange principle, that the strength of the weapon is borrowed from the means employed to thwart it, came to the fore after the First World War, when governments had to consider how to instil gas discipline in the civilian populations they assumed would be the target of gas attacks in future hostilities. Although it is the use of gas in the trenches of the First World War that has been most mythologized, it was during the 1930s, during the long build-up to war, that the threat of gas became the focus of the most intense debate and widely disseminated fears. The focus on the likely use of gas against civilian populations and the means necessary to protect them helped both to amplify and, so to speak, retune the fear and horror of gas. It is this period that transmitted the gas phantasm to the late twentieth century.

It is not possible to consider the effects of gas without considering the place of the mask, on which so much depended. In the run-up to the Second World War, when it was assumed that gas would be dropped from aeroplanes on civilian populations, the gas mask became the most familiar reminder that there was no longer a clear and distinct 'front line' for military conflict. Thirty million gas masks were distributed to the population during the Munich Crisis of 1938. When war was declared, gas masks were provided for almost everyone in urban areas, along with detailed instructions for their use.

The peculiarly intimate nature of the threat posed by gas also depends on the extreme familiarity of the technologies of gas, for cooking, heating and still, at the outbreak of the Second World War, for lighting. Gas was, and is, in almost everybody's home. Most remarkably, gas had been introduced in the United States as a humane form of execution. The 'Humane Death Bill', which abolished all other forms of execution in the State of Nevada, was signed by the governor on 28 March 1921 (previously the condemned man had been given a choice between hanging and shooting). The first use of gas for execution in the US was on 8 February 1924. The victim was Gee Jon, a Chinaman convicted of killing a rival tong man. The *New York Times* reported that 'the Chinaman lapsed into unconsciousness after his first breath of the vaporized acid'.[54] The experience of the lethal effects of gas in the First World War may have suggested the introduction of this method, along with the increased popularity of gas asphyxiation as a method of home suicide. Though one court in California has decreed this method of execution to be cruel and unusual, five US states still authorize its use: Arizona, California, Maryland, Missouri and Wyoming. It has been used 31 times in total, eleven since 1976.

Germany had resumed the secret production of poison gas for military purposes in the mid-1930s, drawing on the dominance in industrial chemical technology that it had never lost, even after the sanctions of the Versailles Treaty. It was widely assumed in Britain and the US that, though gas was certain to be used, no new gases were likely to be discovered. In fact, Gerhard Schrader had discovered the first of the new generation of what would come to be known as nerve gases in 1936, when he found that a compound he named 'tabun' caused generalized muscle contraction and spasm. The discovery of sarin, a compound with similar effects, followed in 1938. (Sarin would be the gas used in the terrorist attack on the Tokyo underground system on 20 March 1995.) Hitler had himself been gassed as a soldier in the First World War, and seems to have had a certain reluctance about provoking retaliation through gas attack (it seems that his advisors wrongly assumed that the Allies must also be in possession of nerve gases).

The other major component in the perception of lethal gas is, of course, its use by the Nazis for the extermination of Jews and others in the concentration camps. A bizarre but insistent logic seems to lie behind this use. The use of gas as a means of extermination has a particular kind of cruelty. The very reputation that the weapon had – as annihilating the humanity of the victim and sometimes also morally contaminating the user – was what seemed to make it appropriate to exterminate peoples who were both regarded as subhuman and proved to be so by the very fact that gas was used on them. Gas has often been used for the destruction of animal pests (the smoking out of moles, badgers and foxes, for example) and the experience of the First World War accelerated research into insecticides. One wonders whether the ancient superstition of the *fetor judaicus*, a particular odour supposed to be emitted by Jews, did not play a part in this grotesque act of ethnic fumigation. The cult of nudity and fresh air that characterized Nazism gave the questions of space a specifically pneumatic dimension. German *Lebensraum* would be guaranteed by clustering Jews together to stew in each other's foul air in the ghetto. Part of the reason for this attribution may be the Christian disavowal of the particular prominence of the sense of smell, and its association with the divine, in the Hebrew tradition: in Hebrew, the word for spirit, *ruach*, is suggestively close to the word for aroma, *rayach*. The God of the Hebrews is often approached through the odour of sacrifice and incense. When the revolver and the machine gun were replaced by Zyklon B as the means of extermination, the logic seemed to be that the parasitic rats should be subject to their own asphyxiating exhalations. The particular power of gas to remove the humanity of one's opponents survives into more recent uses. When Saddam Hussein's forces used the nerve gas tabun at Basra on 17 March 1984, one Iraqi general said defiantly: 'If you gave me a pesticide to throw at those worms of insects, to make them breathe and become exterminated, I'd use it.'[55]

Gas is an embodiment of the new, disembodied, no-man's-land condition both of our technologies and of our wars. It was assumed during the 1930s that civilian war would be a gas war, since it is in the diffusive nature of gas to extend the whole battlefield: 'there is now no very real distinction between the fighting men and their women folk at home'.[56] If conflicts like the last two Gulf Wars continued to conform to the traditional topology of wars, with enemies ranged against each other in opposing territories, and victory measured by the incursion of one party into the territory held by the other, this may be a secondary or compensating effect of the battle against inchoate *furor*, or war of all against all, evoked by Michel Serres.[57] The war against terrorism is really a war against the new conditions of war, of which the drifting, infiltrating, assassinating nature of gas, always both homely and outlandish, the most intimate of adversaries, is an allegory.

Slum of Fire

But the Second World War would be typified by another kind of air war, in the form of aerial bombardment. Peter Sloterdijk sees this as an extension of the war upon immanence begun by the use of poison gas in 1915, but in many ways air war permitted a return to the glamour of the vast and unoccupied air.[58] In many writers, the rhetoric of the air provided a way to distance and sublimate the experience of war. M. H. Pickering's 'The Fliers' is not untypical of the large number of air force poems produced during the war:

> As they fly on against all winds that blow,
> O'er all the endless, 'fenceless fields of air,'
> Uncharted fields in ever-changing sky,
> What isles and worlds of misty realms are there!
> What wondrous beauty in the clouds they know –
> Beauty not dreamt of by those chained below![59]

The ageing Laurence Binyon also looked to the air for relief and release from the contaminations of the lower world. His 1939 poem 'There Is Still Splendour' begins at ground level:

> O when will life taste clean again? For the air
> Is fouled: the world sees, hears; and each day brings
> Vile fume that would corrupt eternal things,
> Were they corruptible . . .

but cheerfully concludes:

> There is still splendour: the sea tells of it . . .
> And the air tells of it: out of the eye's ken
> Wings range and soar, a symbol of the free,
> In the same cause, outspeeding the swift wind.[60]

All this is in violent contrast to the reality of civilian bombing, which had the extraordinary effect of reversing the positions and conditions of the airborne and the one on the ground. The aeroplane was safely at a distance, in a world in which the logic of perception and parallax remains intact. On the ground, in the fire storms generated during the bombings of Hamburg, Cologne, Dresden and other cities (and which it would undoubtedly have been the object of the Luftwaffe to bring about in London and elsewhere had they had the means at their disposal), a new, previously undescribed meteorology of the surface came about. This is W. G. Sebald's evocation of the effect of the bombing of Hamburg on the night of 27 July 1943:

> At 1.20 a.m., a firestorm of an intensity that no one would ever before have thought possible arose. The fire, now rising 2,000 metres into the sky, snatched oxygen to itself so violently that the air currents reached hurricane force, resonating like mighty organs with all their stops pulled out at once.[61]

Though the effects of the fire storm were a hideous discovery, as were the even more apocalyptic contortions of air and sky of the nuclear bomb, air war itself was much meditated on in the late 1930s. It is hard to recapture the certainty felt during the final years of the 1930s that the coming war would be a war in the air. Civilians and governments feared gas delivered from the air even more than bombs. H. G. Wells's prescient novel *The War In The Air* ends, not with any decisive victory, but rather with air war established as a permanent background condition, as evoked in the conversation between an old man and his nephew :

> 'But why did they start the War?'
> 'They couldn't stop themselves. 'Aving them airships made 'em.'
> 'And 'ow did the War end?'
> 'Lord knows if it's ended, boy', said old Tom. 'Lord knows if it's ended.'[62]

In others, more apocalyptic thoughts were prompted. E. H. Horne wrote in 1937 that the coming air war would fulfil the prophecy of the pouring out of the Seventh Vial into the air in the biblibal book of Revelation (16.17). The principal feature of that prophecy, he wrote, 'is, that *the air* is to become the source of all danger, destruction and misery. But this is *the last sign* given in the whole series of numbered visions, and its fulfilment is to bring Christ back, in power and glory, to settle the confusions of mankind'.[63]

The threat of generalized air war seems to have brought about a simultaneous exposure and saturation: exposure to the threat of a death that could at any moment fall from the air, and saturation, because the air was permeated not just with the palpable (and audible) facts of war, but also with the radiating reports of it. Jane Lewty has shown how sensitive Virginia Woolf became to the broadcasts from the BBC that monopolized the ether as thickly as the invading planes threatened to blot out the sky. Others too responded to this condition in which the air bulged as much with bulletins as with bullets. 'Let streams of sweetest air dissolve the blight / And poison of the News, which every hour / Contaminates the ether', wrote David Gascoyne in his apprehensive poem of 1940, 'Walking at Whitsun'.[64] It was as though the exhausted air were finally collapsing in on itself.

It is a tone that passes across into the work of many poets. At the most uncertain moment of her poetic *Trilogy*, in the final poem of *The Walls Do Not Fall*, HD reads the effect of aerial bombardment as a disorientating denaturing of the air. The bursts of explosions have both thickened the free air into a ponderous fog, as though the bombs dropped through the air were the lowering of air itself, and made the riddled ground as unstable as air. The poem gives the experience of being in midair at ground level:

> *Still the walls do not fall,*
> *I do not know why;*
>
> *there is zrr-hiss,*
> *lightning in a not-known*
>
> *unregistered dimension;*
> *we are powerless,*
>
> *dust and powder fill our lungs*
> *our bodies blunder*
>
> *through doors twisted on hinges,*
> *and the lintels slant*
>
> *cross-wise;*
> *we walk continually*
>
> *on thin air*
> *that thickens to a blind fog*
>
> *then step swiftly aside,*
> *for even the air*
>
> *is independable*
> *thick where it should be fine*

and tenuous
where wings separate and open

and the ether
is heavier than the floor

and the floor sags
like a ship foundering [65]

In the end, the poem talks itself up into a rather pallid, poetic sort of hope: ending with the words '*possibly we will reach haven, / heaven*', with a memory of Hopkins's desire to escape 'the sharp and sided hail' in his poem 'Heaven-Haven'.[66]

T. S. Eliot's evocation of the bombing in 'Little Gidding' of 1942 similarly stresses the suffocation of air, seeming to image the horror of and to anticipate the nausea experienced by the witnesses of the destruction of the World Trade Center – of a solid structure reduced to an impalpable but asphyxiating smell:

Dust in the air suspended
Marks the place where a story ended.
Dust inbreathed was a house –
The walls, the wainscot and the mouse.
The death of hope and despair,
This is the death of air.[67]

The meeting with the 'familiar compound ghost' takes place in the usual Eliot landscape of dead leaves rattling in the 'urban dawn wind' after the air raid.[68] Eliot's hope for a pentecostal 'refining fire' is an attempt to strike a light by turning the exhausted, inturned air further in on itself, forming a kind of vortex of fire: 'When the tongues of flame are in-folded / Into the crowned knot of fire'.[69] Eliot does his best to poeticize the incendiary bomb, as a pentecostal dove, which 'descending breaks the air / With flame of incandescent terror', and offers a recommendation that we choose spiritual over literal immolation:

The only hope, or else despair
Lies in the choice of pyre or pyre –
To be redeemed from fire by fire.[70]

His religious rescue of the elements and especially the element of air would be echoed by the larger number of poets and novelists who wrote about the experience of air-raids. It features in Edith Sitwell's popular poem 'Still Falls the Rain', which bears the epigraph '*The Raids, 1940. Night and Dawn*', and begins:

Still falls the Rain –
Dark as the world of man, black as our loss –
Blind as the nineteen hundred and forty nails
Upon the Cross.[71]

Dylan Thomas's poems about urban bombing exhibit a similar eagerness to cook up sacraments out of the most horrible material actualities. 'Ceremony After a Fire Raid' offers us an alarming confection of Christian and Frazerian resurrection myths: 'Oh seed of sons in the loin of the black husk left.'[72] Liturgy joins ludicrously with Dick Whittington in the voluntaristic urban reconstruction of the end of the poem:

Over the whirling ditch of daybreak
Over the sun's hovel and the slum of fire
And the golden pavements laid out in requiems . . .
Glory glory glory.[73]

'A Refusal to Mourn the Death, By Fire, of A Child in London' pompously but distastefully promises that not until the end of time will the poet let slip a whisper of mourning for 'The majesty and burning of the child's death.'[74] The best of Thomas's blitz poems is the more prosaic 'Among Those Killed in the Dawn Raid Was a Man Aged A Hundred'. The octet balances irony and ordinariness in an effective way, though the shift from indicative to imperative in the sixth line suggests some ominous rhetorical limbering-up:

When the morning was waking over the war
He put on his clothes and stepped out and he died,
The locks yawned loose and a blast blew them wide,
He dropped where he loved on the burst pavement stone
And the funeral grains of the slaughtered floor.
Tell his street on its back he stopped a sun
And the craters of his eyes grew springshoots and fire
When all the keys shot from the locks, and rang.[75]

Sure enough, the sestet of the poem flogs itself into a furious slather of symbols that help immunize us from touch and taste and smell:

Dig no more for the chains of his grey-haired heart.
The heavenly ambulance drawn by a wound
Assembling waits for the spade's ring on the cage.
O keep his bones away from the common cart,
The morning is flying on the wings of his age
And a hundred storks perch on the sun's right hand.[76]

This religiosity is gloriously anticipated and spiked by Henry Reed's parody of T. S. Eliot, wickedly entitled 'Chard Whitlow: (Mr Eliot's Sunday Morning Service)'. The title is suggested by *Ash Wednesday* and *Burnt Norton*, but when it appeared as winner of a *New Statesman* competition in May 1941 it seemed to anticipate uncannily the sermonic wranglings over the Blitz in 'Little Gidding', which Eliot seems not to have begun drafting until the following month (published 1942):

As we get older we do not get any younger.
Seasons return, and today I am fifty-five,
And this time last year I was fifty-four,
And this time next year I shall be sixty-two.

And I cannot say I should care (to speak for myself)
To see my time over again – if you can call it time,
Fidgeting uneasily under a draughty stair,
Or counting sleepless nights in the crowded Tube.

There are certain precautions – though none of them very reliable –
Against the blast from bombs, or the flying splinter,
But not against the blast from Heaven, *vento dei venti*,
The wind within a wind, unable to speak for wind;
And the frigid burnings of purgatory will not be touched.[77]

A very different response to the death of the air occurs in Paul Celan's 'Todesfuge' ('Death Fugue'), written in 1944. In this poem, Celan creates a kind of roundelay of recurrent, slightly varying fragmentary phrases evoking the experience of the death camps. The poem seems to conjure a German commandant who sits at night in his quarters writing lyrical lines to his beloved Margarete and then comes out to order the inmates under his command to play, dance and dig graves, possibly their own. One series of variations in the poem weighs the two fates of burial and burning:

wir schaufeln ein Grab in den Lüften da liegt man nicht eng . . .
er ruft streicht dunkler die Geigen steigt ihr als Rauch in die Luft . . .
er hetzt seine Rüden auf uns er schenkt uns ein Grab in der Luft

we shovel a grave in the breezes you can stretch out there . . .
he calls out more darkly now stroke your strings you will lift as smoke into the air . . .
he sets his dogs on us he grants us a grave in the air[78]

The freedom sardonically 'granted' in death is a freedom simply to dissipate as smoke, the release of a non-being that is the most toxic gift. The choice is concentration or dispersal. Celan knows well the importance of the visible location of the body for orthodox Judaism, for whom dying into air, or fire, or ash involves the *Antigone*-like horror of eternal diaspora. The music that Celan so scandalously composes defies and survives its own irony. The poem austerely refuses the nursery-rhyme consolations of John Pudney's famous elegy 'For Johnny', which was written in 1941 but became well known through its appearance in 1945 in Anthony Asquith's film starring Michael Redgrave, *The Way to the Stars* (*Johnny in the Clouds* in the US): 'Do not despair/For Johnny-head-in-air; / He sleeps as sound / As Johnny underground'.[79] The poem was much quoted in the aftermath to the Challenger space shuttle disaster. Nevertheless, as in so many other poems, Celan is attempting to occupy the place and voice of the vanished, of those who are no longer there – as he will again in a poem in which he seems to revisit the site of a camp, to find only *Engführung*, ('straitening') – a word that ironically draws out the 'eng' of the previous poem – where *keine / Rauchseele steigt* ('No smoke soul rises').[80] In thinking through, thinking on the air, Celan seems to be thinking of poetry itself, which is so identified with this element. Roy Fuller voices a similar scepticism about the rich air of poetry in his 'Soliloquy in An Air Raid' of 1942: 'The air, as welcome as morphia, / This "rich, ambiguous, aesthetic air" / Which now I breathe, is an effective diet / Only for actors'.[81]

The grotesqueness of the idea of a people gone up in smoke, disappeared as utterly as this into not quite thin air, is brought out in Martin Amis's *Time's Arrow,* which follows the life of a German concentration camp doctor who is living backwards. His whole life is lived like the inverted explosions he describes in the camps – 'Velocity and fusion sucking up the shocked air'.[82] The narrator imagines a kind of alchemy in which smoke is baked back into flesh, and bodies are snatched from the sky-smearing exhaust that time has made of them:

> Not for its elegance did I come to love the evening sky, hellish red with the gathering souls. Creation is easy. Also ugly. *Hier is kein warum*. Here there is no why. Here there is no when, no how, no where. Our preternatural purpose? To dream a race. To make a people from the weather. From thunder and from lightning. With gas, with electricity, with shit, with fire.[83]

As we move backwards in time from the gassings and incinerations to the experiments in various forms of lethal injection, Amis gives us an image of creation through the removal rather than the inspiration of the breath of life:

> As well as prussic acid and sodium evipan I now extract benzene, gasoline, kerosene and air. Yes, air! Human beings want to be

> alive. Twenty cubic centimetres of air – twenty cubic centimetres of nothing – is all you need to make the difference. So nobody thanks me as, with a hypodermic almost the size of a trombone and my right foot firmly stamped on a patient's chest, I continue to prosecute the war against nothing and air.[84]

Exhaust

Amis's work makes wincingly literal the association between waste and death that has haunted meditations on the relations between death and air. To be shuffled off into the air is to be turned to shit. Increasingly, the idea of death into air has suggested blowback, in the idea of a death that falls from the air, the return to us of the death we are dealing, as carelessly as pissing into the wind, to the air itself. And yet the word we use to describe the contemporary condition of the air, fouled by our inhabitation of it, is an ancient one, which reminds us of the fate of Polyneices: pollution.

The history of air pollution and efforts to combat it has been well-served. There are histories of the growth of different polluting influences, from industrialism to the automobile. There are also studies of the growth of environmentalist pressure groups and legal frameworks for the control of aerial effluents like smoke and exhaust fumes. The cultural phenomenology, or interior life, of these developments has been attended to in much less detail. How has the *idea of air* changed as a result of the increased awareness of the sensitivity and the fragility of the atmosphere?

Most creatures deal with their waste products in two ways. First of all, they discharge, or put their wastes from them. More importantly, they themselves move away from their wastes. The nomad does not merely put waste aside – he puts it behind, in the past, which it then becomes. As human beings settle, cultivating rather than grazing the resources, two alternative movements are set up. On the one hand, there must be a concentration, an influx of resources – fields must be irrigated, workers brought together, food stocks built up. There must be synchronization, not just of human affairs with natural, but also of human beings with themselves, in regular work patterns. At the same time, waste products that might previously have been left behind must be mobilized. As human beings become sedentary, their goods and wastes must be got on the move. Nomadic peoples are themselves fluent, fluctuant; sedentary peoples must create and manage flow, the itineraries of input and output, around them. Hence the importance of rivers, by which many early settlements are formed: rivers bring resources in and take them away, creating a movement through and around spaces of human habitation. But rivers may gradually prove inadequate to cope with the volumes of waste that cities produce, as happened during the years of London's expansion in the nineteenth century. The creation of a sewage system necessitated a new form of map, which had not previously existed, or been necessary – a detailed map of gradients that would ensure that sewage could flow out of the city, at the right speeds and in the right directions.

Previously, human temporality had been formed by itinerary and circulation. Space was temporal (the hills in summer, the plains in winter). The establishment of settlements and cities spatialized time. Time had to be managed, in a series of flows inwards and outwards.

Waste of Space

This active concern with expulsion means that sedentary human beings are forced into more and more intimate relation to their wastes, in the very intensity of their efforts to drive them away. The fortunes of the word 'waste' itself may help to dramatize this. 'Waste' derives from Latin *vastus*, which means empty, wild, uninhabited, and perhaps uninhabitable. It referred to wilderness and desert and, as its afterlife in the word 'vast' suggests, its principal component is the idea of immeasurability or formlessness. The 'vast' is that which human beings cannot encompass or inhabit, in fact or imagination. Already in Latin, the word *vastare* suggested the possibility that human beings could themselves lay waste, could 'devastate'. Gradually, 'waste land' came to mean, not mere unoccupied vastness, but land that had been exhausted, used up, rendered unfit for human habitation. Once, it seems, to devastate meant to confer vastness, to subtract a space from measurable space. We have devised other forms of laying waste, which result in confinement, congestion.

For a mobile people, waste is the past. It no longer exists. A turd is exactly equivalent to an uttered word – no sooner has it risen to the lips than it is irretrievably gone. Once settlements and civilizations arise, and records begin to be kept, once peoples begin to keep time, litter and literature together ensure and require a continuing relation to one's past. Perhaps this is the opposite of the situation endured by Benjamin's Angel of History:

> His eyes are staring, his mouth is open, his wings are spread. This is how one pictures the angel of history. His face is turned toward the past. Where we perceive a chain of events, he sees one single catastrophe which keeps piling wreckage and hurls it in front of his feet. The angel would like to stay, awaken the dead, and make whole what has been smashed. But a storm is blowing in from Paradise; it has got caught in his wings with such a violence that the angel can no longer close them. The storm irresistibly propels him into the future to which his back is turned, while the pile of debris before him grows skyward. This storm is what we call progress.[85]

Modernity must always confront the fear that the propelling gale will flag, or worse still, change its course, blowing backwards from the future, and stranding it in the midst of its own asphyxiating waste

One may suppose that many animals suffer from agoraphobia, from a sense of the pressing, exposing, terrifying emptiness of the world. Most animals have a strong instinct for the particular niches or environments in which they exist and on which they depend: mice, spiders, frogs, all depend on darkness and cover, and very few animals, not even the birds of the air, are able to make a living in the open. Animals presumably do not have a conception of the world, do not have a way of including what excludes them. Human beings, by contrast, whose architecture and culture seem agoraphobic, designed to keep vastness at bay, nevertheless have a relation to vastness, have a conception of the inconceivable, the infinite. They can make of the vast a niche.

The hugeness of the mountain or the plain, stretching out beyond the eye, or the vastness of the ocean, have traditionally stood as synecdoches of the illimitable. But, more than earth or sea, both of which it has long been possible for human beings to traverse, it is the element of air that has stood for immeasurablility. For most of human history, the air has been on the one hand intimately present to human beings and, in the form of the weather, actively influential in human affairs while yet being unknowable and unencompassable. This is because the air is more than a mere element. The air is a quality, a dimension, rather than a mere substance. Where the other elements occupy space, air has habitually been thought of as space itself, the space within which the other elements are disposed. As such, more than the earth or the sea, the air has been seen as the inexhaustible, a pure gratuity.

Luce Irigaray has attempted to reinstate this tradition of thinking in her anamnesis of air, asking 'Is there a dwelling more vast, more spacious, or even more generally peaceful than that of air? . . . No other element is to this extent opening itself . . . were there not an unlimited and always irritatingly excessive resource of air, the open expanse would not take place'.[86] The air is vast, in both senses, then: it is immense, and it is waste, never fully inhabitable by man. The air is a nothing, an out of sight and out of mind. Air does not present itself – 'Air never takes place in the mode of an "entry into presence" . . . in air [the philosopher] does not come up against a being or a thing'.[87] This enables air to be thought of as a universal solvent, the origin and destination of all things. The air thus can become an edgeless, bottomless receptacle for waste.

Time is of the essence in waste management. Aeroplanes and trains used to dribble their wastes unconsciously as they went along, in real time, like birds or cattle. All one had to remember was not to flush the toilet in a station, which was equivalent to fouling one's own nest. But to excrete in transit or at speed was to return to the mode of the nomad, who risks no contamination from his wastes, because he leaves them behind him, in the sand, on the road, in the past. One used to be able to look down through the bottom of the lavatory pan and see the railway track whizzing along underneath. The ideal of shitting into air is to be found enacted early on in the location of privies in castles, which tended to be on the outside corners, in order to allow the wastes to drop straight into the moat below. The ideal location of the privy between the earth and sky made it an appropriate place for contemplation and prayer.

Nowadays, train and aeroplane toilets do not simply trail human waste behind them in inglorious clouds. Rather, they gather and conserve wastes in sealed chemical cisterns for later disposal. The wastes are thereby taken out of time, just as the journey itself is an intermission in time. And yet the manner in which these wastes are removed intimates the survival of an earlier belief in the air's infinite powers to consume and dissolve. There seems to be no water involved in the vacuum-driven flushing of a toilet on a plane. Rather, the excremental products are gulped by a sudden, violent spasm of air (strong enough to seize a wallet, we are assured), as though just for a moment a vent had been open to the freezing, howling, healing waste outside. We may know rationally that the human waste will form part of the aircraft's cargo, rendering it a giant pooper-scooper, which must be sluiced out on arrival at its destination. We have all of us, after all, heard the stories of windshields being shattered by strange blue lumps of ice falling out of the sky underneath flight paths. But our fantasy is fed that our feeble trickles have been plucked away by the jet stream.

The air has been the guarantee of the non-historical temporality of the earth – a temporality founded on fluctuations and recurrences without direction or purpose. The air has time only in one of the two senses in which the French word *temps* can be used – weather rather than history. This sense of the ahistorical temporality of the elements provided the limitless resource on which human accumulation through history has been built. Now, that relation seems to be failing. To recall Michael Serres's remark, not only has history entered nature, but nature has entered history. In a special application of that principle, the element of air, the bearer of the weather that has always embodied pure fluctuation and cyclicity as against the irreversible linearities of history, has crossed over: now, it seems, the air does have a history, and one in which we are implicated.

In fact, Irigaray's female dream of infinitude is the metaphysics of air, not Heidegger's alleged encrusting reduction of it. The metaphysics of air makes air bear the burden of the illimitable, the inexhaustible. Confronted with the illimitable and the inexhaustible, man will always see it as a toilet. Like the ocean, the atmosphere has functioned for centuries as a vast oubliette, into which noxious fumes and vapours could be discharged, in the certainty that they would dissipate. The problem, not to say the disaster, is not the 'male' finitizing of the air, of which Irigaray accuses Heidegger and others, but the infinitizing of the air, the belief in the air as a horn of plenty, a bottomless fund of vastness, emptiness, openness, which can never be overdrawn. The air is not so much a way of having your cake and eating it, as of making your waste and never having to see it again.

The loss of belief in the powers of air to vaporize and dissipate waste is part of that finitizing of the air that has taken place from the late eighteenth century onwards – the loss of belief in the air as the abode of the endless. Air has come down to earth. It has become a quantity rather than a dimension. Rather than extending beyond sight, knowledge and belief, merging into the celestial ether of the upper air, the air has come to be seen as no more than a thin and increasingly patchy rind of breathable gas clinging wispily to the surface of the planet.

The word exhaust derives from *exhaurire*, to draw out or draw off. This was its primary meaning up until the beginning of the seventeenth century. Exhausting was the word used for medicinal drawing of blood, for example. The idea of giving or being drawn off could suggest abundance, as in a sermon preached by Lancelot Andrewes on Easter Day 1606, which refers to Christ's blood 'the vertue and vigor whereof, doth still continue as a fountaine in exhaust, never dry; but flowing still as fresh, as the very first day, His side was first opened'.[88] The exhaustion of Christ's blood is the sign that '*His goings out are everlasting*'.[89] Only by interpretation did exhaustion come to be extended from the idea of drawing off to that of emptying out. In a later sermon, preached by Nathaniel Heywood, we are told that 'tis not enough to have Christ, unless you use him . . . so we must be always drawing, and drinking, and deriving good from Christ, as the branch makes the best of the root, and draws from it, as though it would exhaust all its life and vertue'.[90]

Van Helmont saw in the universal and ceaseless process of exhaustion the secret of death itself:

> unless the blood, yea the very sanies or purulent effluxions from Wounds and Ulcers, the Urine, and that subtle effluvium, which by insensible transpiration evaporates through the pores of the skin, did continually exhaust, and carry with them some part of the vital spirit; and unless these had also some participation of vitality, and conspiracy with the whole body, after their remove from the whole concretum: Undoubtedly the life of man could not be so short. For indeed this is the cause of our intestine calamity, and that principle of death we carry about us, ambuscadoed in the very principles of life.[91]

Thus the object of exhausting came gradually to refer, not to that which was drawn off, but rather the receptacle from which it was drawn off – in the many references to various forms of 'exhausted receiver' in Boyle's writings, for example. At the end of the nineteenth century the nominal form of the word 'exhaust' started to refer, not to the action of drawing out or emptying, but to the product of the process. Despite the chimneys that sprouted up across Britain during the nineteenth century, it was not until the coming of the automobile that the gases vented from exhaust pipes came regularly to be known as 'exhaust'. Exhaust therefore becomes a kind of diffused negativity. Rather than drawing out the air, or the life-principle it embodied, the air itself was rendered exhausted.

After nearly 70 years of debate and attempted legislation, from the first Act to try to mitigate the effects of urban smoke in 1821, through to the Alkali Acts of the 1870s, which attempted to regulate sulphurous and other less visible emissions, the 1890s saw a succession of intense fogs and smogs, probably caused by a mixture of environmental conditions and domestic and industrial emissions.

The late nineteenth century cultivated this ambivalence towards exhaust and aerial waste with a curious intensity and versatility. The smokes and fogs of London were seen as much as an enchantment as a curse. Even Dickens, who took fog as the image of a benighted condition of obfuscation and paralysis in public life, praised the fogs of London. Sara Jeannette Duncan, a Canadian novelist who visited London, even found something to admire in its smell:

> There was the smell to begin with . . . always more pronounced in the heart of the City, than in Kensington for instance. It was no special odour or collection of odours that could be distinguished – it was a rather abstract smell – and yet it gave a kind of solidity and nutriment to the air, and made you feel as if your lungs digested it. There was comfort and support and satisfaction in that smell.[92]

As we saw in chapter Six, fog is also bound up with a growing suspicion of transparency and attraction towards obscurity and blur among the artists and writers who partook of 'nebular modernism', and focused on the rich ambivalence of mist, fog and haze, rather than burning clarity and transparency. The ambivalence towards air pollution is registered in a number of apocalyptic fictions having as their subjects the devastation of the air. One of the earliest to appear was Robert Barr's story 'The Doom of London' (1892). The story is told by an old man, reflecting back on the 1890s in which Barr wrote his story from the mid twentieth century. Both gas and electricity have been rendered obsolete by the discovery of 'vibratory ether', which provides all the necessary power.[93] The narrator tells the story of a lethal fog that gathers one day over London, formed from a combination of water vapour rising from the ground and smoke belched out from domestic chimneys; the latter pressing down upon the former have prevented both the light and the heat of the sun from getting through. The narrator is approached by an American trying to persuade him to adopt a portable oxygen-dispenser. He declines him, but is made a gift of the dispenser anyway. The fog comes, getting thicker and thicker: 'We were, although we did not know it, under an air-proof canopy, and were slowly but surely exhausting the life-giving oxygen around us, and replacing it by poisonous carbonic acid gas.'[94] Once the oxygen in the air begins to fail, Londoners expire in their millions. Only the narrator survives, in the 'oasis of oxygen' furnished by his acclimatizing machine.[95] He eventually escapes London on a train, sharing the machine with its driver, though by now all the passengers on the train are dead. The fate of London is regarded by the inhabitants of the mid-twentieth century as a purgation.

In Arthur Conan Doyle's *The Poison Belt*, it is not the build-up of industrial and domestic wastes that represents the danger, but a belt of poison gas in space through which the earth itself is passing. The story, narrated by a young journalist called Malone, describes the last day and night of life on earth, which he spends in the company of Professor Geoffrey Challenger, and the other members of the band of adventurers who first appeared in Doyle's story *The Lost World*. The very possibility of the story rests upon a central incoherence. The belt of interstellar poison is said to consist of a disturbance in the ether. The four adventurers survive the passage through the poison belt when the rest of the planet is exterminated because they remember to bring with them to Geoffrey Challenger's house in Sussex cylinders of oxygen. One of the band, the sceptical Summerlee, objects to the narrative *donné* that Doyle has devised to keep them alive and narratable: '"But what can oxygen effect in the face of a poisoning of the ether? There is not a greater difference in quality between a brick-bat and a gas than there is between oxygen and ether. They are different planes of matter. They cannot impinge upon one another."'[96]

The other peculiarity of this poisonous ether is that it seems to behave much more like a flood of water than a cosmic gas. Those on high ground are affected much later than those in the plains, and women, children and inferior peoples succumb to the effects of the poison much more readily than sturdy, stiff-backed Englishmen like Challenger and his indomitable party. The first signs of the effects of the cosmic poison are felt among 'the indigenous races of Sumatra', who seem to respond more quickly to the cosmic conditions than 'the more complex peoples of Europe'.[97] Although death for all seems certain in the end – 'painless but inevitable – death for young and old, for weak and strong, for rich and poor, without hope or possibility of escape' – there is a differential topography of annihilation, partly because the preliminary effects of the poison involve disturbances of the rational faculties.[98] The telephone in Challenger's house brings news of the slow spread from south to north:

> The great shadow was creeping up from the south like a rising tide of death. Egypt had gone through its delirium and was now comatose. Spain and Portugal, after a wild frenzy in which the Clericals and the Anarchists had fought most desperately, were now fallen silent. No cable messages were received any longer from South America. In North America the Southern States, after some terrible racial rioting, had succumbed to the poison. North of Maryland the effect was not yet marked, and in Canada it was hardly perceptible. Belgium, Holland, and Denmark had each in turn been affected.[99]

This is interesting despite, or in fact because of, its implausibility. 'Ether', which for so many was imagined as a kind of idealization of the air – air refined and attenuated to the ultimate degree, and thus spiritualized – here becomes an annihilating, asphyxiating agent. The poison belt of malignant ether, into which the earth passes purely by chance, provides an image of the air both desublimated and moralized. The uncertainty as to whether ether can be thought of as material or immaterial carries this ambivalence towards the air. Doyle is attempting to hold together the two dimensions of the air – the air as infinite space, maximal communicability, and the air as finitely subject to the laws of spatial distribution. Ether represents the infinite space that has been slowed and thickened into place.

In fact, it turns out that the great extermination is merely temporary, for gas acts as an anaesthetic inducing catalepsy, rather than actually killing its victims. This means that the end of the world is followed by a great awakening, which is spiritual as well as physical, and has longer-term effects, in the turn away from a culture of excess and consumption:

> Surely we are agreed that the more sober and restrained pleasures of the present are deeper as well as wiser than the noisy, foolish hustle which passed so often for enjoyment in the days of old – days so recent and yet already so inconceivable. Those empty lives which were wasted in aimless visiting and being visited, in the worry of great and unnecessary households, in the arranging and eating of

> elaborate and tedious meals, have now found rest and health in the reading, the music, the gentle family communion which comes from a simpler and saner division of their time. With greater health and greater pleasure they are richer than before, even after they have paid those increased contributions to the common fund which have so raised the standard of life in these islands.[100]

Thus, the poison turns out to be sanative and restorative. Doyle fantasizes a world of proportion and limit, brought about by the accident of a cosmic encounter.

As we have come to draw the air into our sphere, even to come ourselves to occupy the mid-air in which our mass communications and movements take place, we have indeed, as Irigaray charges, forgotten the dream of the infinitude of the air. But that dream is fatally implicated in the history of man's efforts to put waste definitively out of sight and out of mind in the all-encompassing and all-consuming invisibility of the air. The philosophical discussion of waste has, since Georges Bataille, been tempted by the pull of the unaccountable, of pure gratuity or excess. The compromising of the idea of air suggests that we should suspect and resist the return of infinitude, the modernist yearning to escape place, limit and economy. By contrast, this claustration, this implicatedness, this accounting for and economizing on air, may be the only way in which we can temper our savagely negligent dream of the air as a vastness that we can never lay waste.

On the one hand, our modernity is characterized by a growing sense of the rapture of dying by, or dying into air. On the other hand, there is a growing fear of the annihilation of the nothing that the air has always been. On the one hand, a grave in the air wherein one may lie unconfined: on the other, the death of air. We no longer live only in fear of annihilation; from the nineteenth century onwards, the fear began to grow of indefinite survival. We fear the annihilation of the infinite, or the medium that stood for it, the air, into which we might diffuse. Alongside and threaded through the fear of not being there is the fear of not not being there; the fear of living on in a state of radioactive decay, of undying by degrees, half-life by asymptotic half-life.

The air, the (un)place of miracle, vision, epiphany, apocalypse and apotheosis for previous epochs, has been disenchanted in our epoch of air. Air has largely lost its qualities, its values and powers, has become 'thin air' – uniform, abstract, horizontal, indifferent, insipid, qualityless. Air is the in-between, the milieu or mid-place, space, the media age's universal medium. Air is no longer the beyond, but the between. It has also been colonized. The reduction of air to the condition of space, medium or interval – to airtime – has been produced by, and itself produces, a saturation of the air by the human. The air, which had always before represented the enclosing openness that gave us our location, the space that subtended our place, has become inundated with our traffic, our signals, our detritus. As Vladimir proclaims in *Waiting for Godot*, 'the air is full of our cries'.[101] Philip Pullman exulted to see the sky 'full of dad'; but the air has been exhausted of its invigorating vacuity by being filled with us. Without ever being able to habituate ourselves to the air, or fully to inhabit it, we have nevertheless brought the air under occupation. The air has been finitized, unothered. But it is precisely this that constitutes the menace, the mortification of the air.

PART THREE

Absolute Levity

Richard Dighton, 'One of the Advantages of Gas Over Oil', an optimistic sentiment attached to an etching of a woman blown off her feet by an explosion in a shop selling oil-lamps (London, 1821).

nine

Air's Exaltation

The idea of the material imagination that has guided this book is that human beings take themselves for the world, by taking the world for themselves. To be sure, this world, this material world of substances and processes, is not simply available for human assimilation. It must be imagined, and imagined in material ways, because all the ways in which mind and world are imagined involve material analogies ('assimilation', for example). I have tried in these pages to substantiate the hopeful hunch that there is a particularly strong affinity between acts of mind and ways of imagining those acts of mind on the one hand, and the imagination of air on the other – if only because the act of imagining has traditionally been so airy.

But the air is not all arias, auras and aromas. Our long and compound dream of air also includes danger, death, decision and unpredictability. All through this book I have also assumed a distinction between the active and passive dimensions of air. As the domain of dream and spirit, the air is pure medium, that 'next-to-nothing' through which human desire reaches. In its passive or hospitable aspect, the air is dimension and extension themselves. But though invisible, the air is far from being impalpable; though insubstantial, it is far from powerless. From Robert Boyle onwards, modern scientific enquiry has set itself to explore and understand the powers of the air, conceived now not as the gods and demons of the wind, and the 'influences' breathed from the stars, but as mechanical forms, forces and processes.

Of course, the wind, then as now, was a capricious and unreliable force. But there was an even more capricious force, which also began to be known and used in the late medieval period. Previously latent and occasional in the eras of the 'open air', it had ever more to be reckoned with in the era of investigation and exploitation that was brought about by the many ways of enclosing the air, thereby concentrating its force. This is the force of explosion. An explosion always seems to be some convulsion of air. An underground or undersea explosion is a thwarted, subjunctive explosion, an explosion that has not quite happened. Of course, no explosion can happen without some kind of confinement or compression, and explosions where the force is contained within some confined space – in mines, or, as terrorists know, crowded places (stations, buses, Tube trains) – are particularly deadly. But the mark of the explosion is its sudden, catastrophic out-ering, its violent throwing out into open space of pure impulse. Samuel Sturmy wrote in his *Mariners Magazine* (1669) of the effects of gunpowder in propelling a charge, that 'The Shot is driven forth . . . by the Air's exaltation, or Wind, caused through the Salt-Peter'.[1] Ex-altation means a lifting up: here it is as though Sturmy wanted us to hear in the word an 'ex-saltation', too, a leaping forth. For us, in an era in which most armaments – and accidents – have the capacity not just to mangle but to vaporize their victims, the explosion stands for the power to blast things not merely out into the air, but also into the condition of thin air, the power of air violently to take things into its condition.

The uses of explosion are not confined to their obvious physical applications – fireworks and spectacle, military uses, mining, demolition and, eventually, transport. The experience of explosion enters and alters language, and the work of thought it allows. The idea of explosion has become an indispensable analogue for a huge range of processes, physical, natural and social. Finally, there is an elective affinity between explosion and the one truly distinctive art-form of the twentieth century, moving images, the form of which is inextricably entangled with explosion. Explosion becomes a way of seeing, a logic of syncopation that connects nature and humanity. Explosion represents the most important form of our newly unfolding involution with air.

It's a Blast

From the very beginning, explosion has constituted a dangerous pleasure: always more than a physical event, it has been inflected by and given an accent to desire.

It is not clear when juvenile male hominids first began to take the pleasure they continue to take in plosive play, or which were the first cultures to organize their laughter systematically around farts and raspberries, burps and barfs. But it is a cheerful thought that the principal function for gunpowder, the substance that would be the sole and undisputed means for effecting and controlling explosions for about 1,000 years, may originally have been festive rather than military. It is thought that gunpowder was discovered by Chinese alchemists around the ninth century. Gunpowder has three constituents: saltpetre (potassium nitrate), sulphur and charcoal. The chemistry is straightforward enough: the sulphur provides a source of easy ignition and lowers the burning point of the mixture; the saltpetre furnishes a rich source of oxygen; and the charcoal provides the fuel for the reaction. When the mixture is ignited, it produces a large amount of hot gas very quickly. When the mixture is contained rather than ignited in the open air, the pressure of the rapidly expanding gas may be enough to shatter the container, expelling the fragments at great and, of course, injurious force. The principal use for gunpowder in China seems to have been in pyrotechnics: the mixture would be packed inside bamboo canes which burst with a loud and exciting crack, or would propel a capsule across the ground or into the air. Marco Polo brought back stories of these arduous entertainments, writing that '[they] burn with such a dreadful noise that it can be heard 10 miles at night, and anyone not used to it can easily get into a swoon or even die'.[2] As gunpowder began to be brought into Europe by Arab traders, pyrotechnic uses were high on the list of favoured applications. Pyrotechnics were important in the arts of theatre and other forms of public display, and an important component of the natural magic that formed a crucible for natural and experimental sciences of the seventeenth century. Handbooks of pyrotechnic construction continued to do a brisk trade into the eighteenth century, setting out methods that had not changed very much for centuries for how to make fireworks '[s]uch as operate in the Ayer, as Rockets, Serpents, Raining Fier, Stars, Petards, Dragons, Fier-drakes, Fiends, Gyronels, Fier-wheels, or Balloons', as well as fireworks that burned and burst on the ground and on water.[3]

There is an obvious link between explosion and sexual reproduction. Nature employs many forms of explosion to piggyback on chance dispersal. As a result, explosion has come to assume the meaning of radical, renewing abundance. When seedpods are designed to explode their cargoes, the process is known as dehiscence, a word that has found particular favour with Samuel Beckett and Jacques Derrida. Among many plants that distribute their seeds by this mode of 'active mechanical dispersal' (as opposed to the passive mechanical dispersal that employs hooks and barbs to allow the seed to be carried away), are *Hamamelis* or witch hazel, which has pods with two compartments, each containing a seed, which are popped open by the warmth of the sun, gorse (*Ulex europaeus*), many varieties of Busy Lizzie (*Impatiens*), and broom (*Planta genêt*), the plant from which the Plantagenets derived their name, which releases its pip-like seeds through a mechanism whereby the two halves of the pod twist away from each other. The brooms (*Genista*) have taken over large areas of California, and the sound of their seedpods popping en masse is remarkable. Julian Barnes tells of a dream recounted by Alphonse Daudet to Edmond de Goncourt: 'he was walking through a field of broom. All around him there was the soft background noise of seed-pods exploding. Our lives, he had concluded, amount to no more than this: just a quiet crackle of popping pods.'[4]

There is also an obvious bridge between reproductive and aggressive explosion to be found in the 'grenade', an early word for a pomegranate, literally, seed-apple, because of its abundance of seeds (it is also known as 'forest tomatoes'). The pomegranate is a symbol of fertility and abundance for Greek Orthodox Christianity, with pomegranates being traditionally broken on the ground for luck at Greek weddings. The OED records the first appearance of the word 'granade' to mean a small explosive or incendiary shell in William Garrard's *Art of Warre*.[5] A 1690 work on ballistics spelled out the lesson that 'As a Granarium keeps Corn for the Preservation of the Life of Man, so these

Filling a grenade, from Francis Malthus, *A Treatise of Artificial Fire-works Both for Warres and Recreation . . .*, trans. Thomas Cecil (London, 1629).

Granariums (corruptly called Granadoes) are filled with Corns of Fire for the Destruction of Mankind'.[6] It went on to specify its construction: 'A Granade is a hallow Sphere of Iron (as we may so call it) fill'd with Corn-Powder, with a Fusee to fire the dry Powder to break the Shell when it arrives to the designed Object'.[7] Francis Malthus's *Treatise of Artificial Fire-works* of 1629 gave fuller instructions, explaining that 'All granads being made to breake, ought to be composed of the most brickle mettle that may be found, as of brasse, adding the third part of Tin to it', and specifying that it should be filled with fine gunpowder, have a screw applied to the top, with holes in it for fuses, and then be covered with pitch in order to preserve it.[8] Malthus guaranteed his readers that 'these sorts of granads worke great effects falling upon houses; they cast downe the walles and coverings, likewise they worke wonderous operations amongst either horsemen or footmen, tearing both man and beast in pieces, sparing nothing'.[9] Given the size of the grenade depicted in his work, one may well believe it. As the century wore on, it became more common for grenades to be thrown by hand, following the kind of technique set out in Joseph Moxon's *Epitome of the Whole Art of War* (1692).[10] The grenade was sufficiently well known in 1675 for Richard Baxter to use it as a metaphor for the explosive effects of the powerful but imperfectly understood word 'soul': 'this one poor word is the Granade or Fire-Ball'.[11] Even earlier, Richard Lovelace's poem warning a fly against immersion in burnt claret (for which we are to read the allegorical incinerations of love) predicted 'An Icy breast in it betray'd, / Breaks a destructive wild Granade'.[12] Altogether, the association between the explosions of armour and *amour* was well established. In Susanna Centlivre's play *Mar-plot*, Colonel Ravelin lamented of the object of his affections that

> she maintains her Ground too well, there's more danger of my raising the Siege, than her beating the Chamade; she has so many retreats of Pride, Vanity, and Affectation, that without some lucky Accident toss a Granade into the Magazine of her Inclination, there'll be no hopes of the Town.[13]

Although the airy associations of explosions, through expressions like 'blowing up', 'blowing away' and 'blowing out', are clear enough in the early explosive era, the words 'explode' and 'exploding' are used almost entirely in metaphorical senses throughout the seventeenth century, to mean refuting or discrediting of opinion. This usage becomes clear in a remark of Henry More's, discussing the materialism of Hobbes, whom he describes as 'one who, upon pretence that all the Appearances in the Universe may be resolved into meer Corporeal causes, has with unparallell'd confidence, and not without some wit, derided and exploded all immaterial Substance out of the World'.[14] No chemical process is being asserted here, and More seems confident that there will be no misunderstanding of his meaning. Margaret Cavendish uses the phrase 'the *Exploding of Immaterial substances*' to refer similarly to the denial of a doctrine.[15] Exploding could also mean 'expelling'. The rules of a castaway colony decreed 'That who should defame or speak evil of the Governour, or refuse to come before him upon Summons, should receive a punishment by whipping with Rods, and afterwards be exploded from the society of all the rest of the inhabitants.'[16] Edward Phillips's *New World of English Words* of 1658 defined 'explosion' as 'an exploding, a sleighting, or hissing off from the Stage', and is followed in this by Elisha Coles's *English Dictionary* of 1677, which defines 'to explode' as 'to hiss off the stage', and 'explosion' as 'an exploding, slighting, decrying, hissing, off, &c'.[17] It is perhaps a mark of the ever more palpable experience of the explosion that the word tends to be used progressively in a more literal, rather than a metaphorical sense, or at least that the metaphors are imbued with a more actual experience of explosion.

Fireworks were domesticated in the UK (to mark Guy Fawkes night) and the USA (to mark Independence Day) during the nineteenth century. The increasingly elaborate form of the latter drew the following sardonic commentary from a *New York Times* columnist in 1877:

> For weeks before the Fourth of July the approach of the National Anniversary is heralded by the blowing up of manufacturies of fireworks, and the diffusion of small particles of exploded workmen over miles of startled country . . . Now that fulminate of silver, nitro-glycerine and other violent explosives have been converted into playthings for juvenile patriots, the blowing up of a single small boy may furnish business for a dozen coroners, each of whom may reasonably expect to pick up a finger, an ear, a jack-knife, or other organ of the victim, upon which an entire inquest can lawfully be held. Not long ago a manufactury of nitro-glycerine exploded in New York and scattered finely-comminuted workmen all over the neighbourhood . . . In the midst of life we are also in the midst of fireworks, and no man knows at what moment his ears may be deafened by an explosion and his hat flattened over his eyes by the parabolic descent of some total though mangled stranger.[18]

During the nineteenth century explosions became not only more spectacularly powerful, but also more likely to be the subject of spectacle. On 10 October 1885 a large crowd assembled to witness the blowing up of Flood Rock, a partially submerged hazard to shipping: Hell Gate, the narrow channel around the rock, was cleared for traffic.[19] Explosive demolitions of buildings have also drawn large crowds. Perhaps the most impressive explosive spectacles of the twentieth century were the launches of the Saturn rockets during the 1960s and 1970s. But, no matter how familiar we may have become with them, the pleasures of witnessing explosions persist, and even intensify.

There is no area in which the pleasurable internalization of the effects of explosion is more clearly signalled than in the language of intoxication – in which one is 'bombed', or experiences the 'hit' or 'blast' of a drug (the kind that from the 1960s onwards could be relied upon to 'blow your mind'). Jacob Schmidt explains in *Narcotics Lingo and Lore* that to 'blast Mary Jane to kingdom come' means to smoke marijuana furiously and unremittingly.[20] 'Dynamite' was the name of a particularly powerful kind of heroin.

Of course, the military possibilities of explosive materials were also seen early. Initially, the Chinese used gunpowder as an incendiary or propellant device. Gradually, as it became clear that increasing the proportion of saltpetre in the mix produced more powerful and destructive explosions, the gunpowder began to be encased in balls and canisters that could then be used as explosive devices proper against an enemy. The thirteenth century saw the appearance of the more destructive 'thundercrash bomb', which yielded the first characteristic descriptions of the annihilating effects of explosion: 'the attacking soldiers were all blown to bits, not even a trace being left behind', reads one contemporary account.[21]

The prospect of blowing an antagonist to bits, ultimately reducing him to nothing, or the next to nothing of dusty air, is the extreme form of the explosion fantasy, which we may suppose is nevertheless immanent in it from the very beginning. Where ordinary forms of death-dealing leave the inconvenient reproach of a corpse to be disposed of, the explosion seems to reach further: the one who is blown to bits has been blown out of the course of life and death altogether. Explosion promises a death without trace, memorial, or residue. It is as though the explosion reaches back into history, ensuring not just that its victims are inexistent, but that they will never have been. This is the ultimate reach of the death-drive, a death that not only brings life to an end, but also annuls the very time of its having existed.

Spasmodick Pathologie

The body could be not only the casualty of explosion, but also its seat and source. The deep, centuries-long concern with the pneumatic dimensions of the body – with the effects of winds, vapours and spirits, as they are variously distilled, confined and expelled through the various chambers of the body – encouraged accounts of the body's operation in terms of explosive action. The most influential of these accounts was provided by Thomas Willis, who had a strong interest in such topics as fermentation as well as medicine. He developed elaborate analogies between the action of explosion and the operation of muscular force:

> . . . when this force, only depends on the expansion or rarefaction of the Spirits, seated in the motive part, we can conceive it to be no otherwise, but that the Spirits so expansed or stretched forth, as it were fired, after the manner of gunpowder, to be exploded or thrown out. But we may suppose, that to the Spirituous Saline particles, of the spirits inhabiting the interwoven *fibres* in the muscle, other nitrous-sulphureous particles, of a diverse kind, do come, and grow intimately with them, from the arterous blood, flowing every where within the same fibres. Then as often as the particles of either kinde, as *Nitre* and *Sulphur* combined together, by reason of the instinct of motion brought through the nerves, are moved, as an inkindling of fire, forthwith on the other side bursting forth, or being exploded, they suddenly blow up the Muscle, and from thence cause a most strong drawing together.[22]

That this was a somewhat unfamiliar notion is clear from Willis's apologies for its novelty, and solicitude in providing examples of explosive processes that might be analogous to those in the body:

> If any one shall be displeased at the word *Explosion*, not yet used in Philosophy or Medicine, so that this Spasmodick Pathologie, standing on this *basis*, may seem only *ignoti per ignotius*

> *explicatio*, an explication of unknown things by more unknown things; it will be easy to shew the effect of this kinde of notion, and very many examples, and instances both concerning natural and artificiall things; from the Analogie of whose motions, in an animated body, both regularly and irregularly performed, most apt reasons are to be taken.[23]

The idea that '*[t]he Motion of a Muscle is a certain explosion of the Spirits*'[24] helped to solve the mechanical problem of how the thin and refined animal spirits were able to govern and move a bodily frame that was so much denser and heavier than they were. Willis calls to his aid the French natural philosopher Pierre Gassendi, pondering the problem of how animal spirits can be responsible for moving an elephant, and suggesting that

> the same fiery nature of the soul serves chiefly to this, which, although it be a very little flame, it is able to perform within the body, by its own mobility, the same thing in proportion, that a little flame of Gun-powder does in a Cannon: whilst that it not only drives forth the Bullet, with so much force, but also drives back the whole machine, with so great strength.[25]

This internalization of explosion makes it a continuous endogenous threat, as well as a source of amusement (perhaps, as we have seen, the very *form* of amusement). The human intuition that explosion may be a feature of the body itself chimes with the fact that explosion is so apt to come home to its perpetrators. At the very heart of the idea of the explosion, which after all constitutes an annihilation of space, is the possibility of being hoist with one's own petard, in Hamlet's phrase. Indeed, the very name of that device, used for blowing open gates and doors during sieges, has a physical application, for it derives from the French *pêter*, to fart. Lest we be in any doubt of the bodily associations of Hamlet's petard, we can turn to Cotgrave's French dictionary of 1611, which not only gives us the noun 'petard' and the accompanying verb 'petarder', but also 'petarrade', defined as '*Gunshot of farting; also a horses kicking, winsing, or yerking out*

A petard in operation, from Robert Ward, *Animadversions of Warre* (London, 1639).

behind, accompanied, for the most part, with farting.'[26] The name of the device seems clearly to derive from its design, which was long and trumpet-like, as described by Robert Ward in his *Animadversions of War* (1639), who says that petards are 'fashioned like to a Morter or Saints-Bell somewhat tapered, they are ¼ parts Diameter of their mouthes Calibre at the bottome of their Chase; and in thicknesse of mettall ⅛ parts Diameter at their Breech, and up towards their mouth, lessening in thicknesse'.[27] Ward promised that the force of a petard was such that 'all substantial massie bodyes are dissolved and fittered in peeces'.[28] It is clear from the haste with which the petardier is retiring from the scene in Ward's illustration that he is sensible of the danger of being levitated by his own infernal engine. It may be that the design of the petard made blowback more likely than being lifted into the air (hence, perhaps the name). This had certainly always been a danger with the mines that were placed under the foundations of buildings during sieges, which often employed tunnels; some siege engineers recommended digging a serpentine tunnel to gain access to the enemy's foundations, precisely in order to contain and dissipate the force of the backfire. Hamlet's thoughts actually seem to be more with explosion into air, rather than explosions channelled backward, for a moment after his petard metaphor, he promises 'I will delve one yard below their mines / And blow them at the moon'.[29]

There was another reason why explosions might have been linked with human bodies. The 'nitre' or saltpetre that was known to be an active ingredient in gunpowder was thought to be a kind of vitalizing force, found both in the air and in human bodies. The firework-maker John Bate prefaces his firework recipes with the remark 'The Saltpeter is the Soul, the Sulphur the Life, and the Coales the Body of it'.[30] The desiccated old Lady Tub in Ben Jonson's *Tale of a Tub* (1595–8) is said to be 'such a vessell of fæces: all dry'd earth! / *Terra damnata*, not a drop of salt! / Or Peeter in her! All her Nitre is gone'.[31] 'Petering out' may well contain a reference to saltpetre – and the OED, somewhat mysteriously, also suggests a link with French *péter*, to explode. Saltpetre was used as a medicine, and is in fact particularly effective, like the range of nitrites that include amyl nitrite, in treating asthma and angina, through its action of dilating constricted blood-vessels and bronchial pathways (amyl nitrite tablets are popularly designated with the mildly explosive name of 'poppers'). This relaxant effect seems to have contributed to the reputation of saltpetre as an anaphrodisiac, which according to rumour was added to the food in schools, barracks and other all-male institutions (though this also contradicts its more explosive aspects). Other explosive substances were also used as medicines. 'Aurum fulminans' (gold hydrazide, $AuHNNH_2$) was discovered by alchemists. Thomas Willis and Robert Hooke demonstrated in 1659 that it could be detonated by concussion, making it the first high explosive to be discovered. Its dangers were reported by one luckless eighteenth-century chemist:

> Having about three Drachms of this Medicine lying upon a Balneum Heat, I went to it next Day to see if it was dry, when thinking all things were grown cold, I began to unwrap the paper to view it, when it instantly exploded with a terrible Noise, and exhibited a Flash of Light, which is a Phaenomenon I have not heard taken notice of by others: This Explosion of the thundering Gold hurt both my

> Eyes, occasioning a great Flux of Blood to them: It also gave a great Contusion to my Hand, by striking its Force downwards, which is contrary to the Assertion of others who write upon this Subject, and say, that it only gives its Force upwards, or sideways.[32]

Nitre was itself indirectly a product of human bodies. It occurs naturally in ground where animal or vegetable matter has decomposed; ammonia from urea and faeces will react with bacteria to create deposits in the ground. Its crystals could be found on the walls of caves (hence the alternative name 'salt of rock'), but were also common in stables and latrines. As we saw in chapter One, 'aerial nitre' was the name given to the substance in the air that John Mayow presciently thought supported both combustion and respiration.[33] The demand for saltpetre increased hugely during the sixteenth and seventeenth centuries, especially in expanding and bellicose countries like France, Spain, England and the Netherlands. The Salpêtrière Hospital, the scene of so much medical convulsion, was founded on the site of a nitre-bed and gunpowder factory in Paris, giving it its name. Elizabeth granted licences and monopolies for the production of saltpetre from the middle of the sixteenth century onwards in an effort to remove the dependence on French supplies of gunpowder. In 1625 Charles I empowered saltpetre manufacturers to enter any dwelling house to remove nitrate-containing earth, and in the following year issued a proclamation requiring that all subjects 'shall carefully and constantly keep and preserve all the urine of man during the whole year, and all the stale of beasts which they can save and gather together'.[34] The consolidation of British imperial power from the eighteenth century onwards had a good deal to do with the fact that the East India Company, under Clive of India, was able to establish in 1758 a monopoly over the extensive saltpetre deposits in India and Afghanistan, breaking the control over supplies that had previously rested with the Netherlands.[35]

The contribution of human bodies to explosion takes a more intensified form in the doctrine of spontaneous human combustion, stories of which began to circulate during the eighteenth century. Although spontaneous human combustion was thought to result from the ignition of gases produced from within the human frame, the element of explosion seems to be missing from most accounts, including literary accounts.[36] The narrator of Captain Marryat's *Jacob Faithful* (1834) is launched into independent life after his obese, gin-bibbing mother dies from 'an inflammation of the gases generated from the spirits absorbed into the system'.[37] There is a shriek, a smell of burning, and '[a] strong empyreumatic thick smoke', but no explosion; instead, there is a slow, smoky casserole of the 'unwieldy, bloated mountain of flesh' that was his mother, leaving 'a sort of unctuous pitchy cinder' in the bed.[38] The death of Uncle Macquart in Zola's *Doctor Pascal* (1983) is very similar: 'the little blue flame arose from [his flesh], volatile, flickering like the roving flame on the surface of a bowlful of punch. As yet it was barely taller than a night-light, so gentle and so unstable too, that the faintest puff of air sufficed to displace it.'[39] Similarly in Melville's *Redburn* (1849), a drunken sailor is seen burning up, again with the slow consumption of a flickering flame:

> [T]wo threads of greenish fire, like a forked tongue, darted out between the lips; and in a moment, the cadaverous face was crawled over by a swarm of worm-like flames . . . covered all over with spires and sparkles of flame, that faintly crackled in the silence, the uncovered parts of the body burned before us, precisely like a phosphorescent shark in a midnight sea.'[40]

In *Bleak House*, which seems to ache for an explosion of fresh air to clear out the various kinds of contagion and congestion that afflict the social and material atmosphere of the novel, Krook's spontaneous combustion is a markedly low-octane affair, for he seems to pass straight from his combustibly spirituous condition into soot, slime and ashes, seemingly without the intermediary phase of flame or fire-burst. There is a kind of accessory explosion in the novel, when the Jarndyce case collapses, with 'people . . . streaming out looking flushed and hot, and bringing a quantity of bad air with them'.[41] But, in reality, this is no exothermic release of energy or potential, since the entire case has by now consumed itself in costs, with the result that, like Krook, 'the suit lapses and melts away'.[42]

The exception to all these damp squibs is Charles Brockden Brown's *Wieland*. In the opening chapter, the narrator's father, who has exhausted and inflamed himself with excessive religious devotion, retires to an outdoor temple to pray, watched anxiously by his wife:

> Her eyes were fixed upon the rock; suddenly it was illuminated. A light proceeding from the edifice, made every part of the scene visible. A gleam diffused itself over the intermediate space, and instantly a loud report, like the explosion of a mine, followed.[43]

The father is found scorched and dying, amid a 'cloud impregnated with light'.[44] No explanation is given for this event, which also seems to have no bearing on the strange narrative of ventriloquism, deception and murder that follows.

The Tenth of a Second

The increasingly routine use of explosives in war produced ever-greater danger of accidents. In one of the earliest and most devastating, the explosion of a powder magazine in the city of Limerick on 12 February 1694 killed over 100, with one contemporary account reporting that the entire quay, was 'blown to admiration'.[45] Accidents became more and more common during the nineteenth century, which saw the development of a great number of new, much more powerful explosives. Nitroglycerine was discovered in 1847. Its power, like that of other high explosives, comes not from deflagration, but from detonation: that is, it does not burn through the available material, but spreads its force by means of a shock wave, travelling at speeds close to the speed of sound, the pressure of which detonates the surrounding fuel. Its power was obvious, but its instability, deriving from the fact that it freezes only at 13 degrees centigrade, made it extremely hazardous to manufacture, handle and transport. Under the name glycerol trinitrate, it too, like saltpetre, has medical applications, being used as a vasodilator, and thus being useful in the treatment of angina. There are also reports of it being used in conjunction with condoms to dilate the blood vessels of the penis and thus stimulate erection. However, exposure to nitroglycerine also causes severe headache, though this can be alleviated by long exposure. Workers returning from holiday found that they had lost their habituation to these symptoms, and so there are hair-raising stories of workers carrying round little prophylactic phials of nitroglycerine to keep 'NG head' at bay.

The use of high explosive was essential to the construction industry. It was in regular use, not just in mining and quarrying, for the ores and minerals necessary to the development of industry, but also for blasting through landscapes to build roads and railways. The result was that explosions began to occur more and more regularly, and more often close to centres of population, becoming in the process less and less a feature of the battlefield and more and more widely reported as a feature of civilian life. A new mode, or rhythm of life – and imminent death – seemed to some to have been instituted:

> In older times, men wore out slowly, by labor or by rust; they set about dying deliberately, as they worked their land or managed their daily concernments. But in these days of steam and dynamite our mode of death is sudden, quick and certain, like an explosion or a railway catastrophe; less like the processes of nature than those of man.[46]

An enormous nitroglycerine explosion took place in San Francisco in April 1866. In the following year, Alfred Nobel patented a mixture of nitroglycerine and diatomaceous earth, a chalk-like sediment, which he marketed, initially under the name Nobel's Safety Powder, and then, a little later, as dynamite. Time itself has begun to be apprehended as routinely convulsive.

Different forms and names of explosive multiplied. Litho-fracteur, a patented composition by Engels of Cologne, first came to notice when the Prussians used it in the siege of Paris in 1871. Rackarock, a mixture of potassium chlorate and nitrobenzene, the name of which came from the fact that it was commonly used to blow up rocks that were causing hazardous obstructions, was patented in 1889. It was joined by Roburite, Hellhoffite, Panclastite, Ergite, and many more. But it was dynamite that spread and domesticated the idea of high explosive. This was principally because it became the favoured explosive of the many militant, revolutionary and anarchist groups that sprang up in the last two decades of the nineteenth century. A new figure began to haunt the modern urban imagination: the dynamiter capable of bringing about instant annihilation in the crowded middle of the city. One of the most shocking plots of the 1870s was the explosion of the steamer *Moselle* in the harbour at Bremer on 11 December 1875. A case of dynamite belonging to William King Thomas, or possibly Thomson, from Brooklyn exploded on the quay, killing more than a hundred passengers. Thomas had set his bomb to explode when the ship was in mid-Atlantic, after he had himself disembarked at Southampton. The shocking thing to many was that Thomas appeared to have been intending to sink the ship simply in order to get insurance money for lost articles. One of the more sensational press responses to the outrage was driven to posit a kind of general explosiveness in human nature and history:

> Human nature, according to all that we know of it, never reached so low a level as that before. Compared with such an act, a mere murder is venial, a massacre for religion's sake is respectable. To call it fiendish is to slander the devil. Milton's Satan would have turned such a creature as Thomas out of hell. If there be many Thomases in the world, we may, indeed, not only doubt that human nature is morally better than it was; we may be sure that it is morally very much worse. It is one of the most frightful thoughts that has been presented to the human mind, that possibly there are such men around us; that we meet them in social and commercial intercourse – men who would destroy us and those we love in a moment, in the twinkling of an eye, if they could get gain thereby, and be sure that they would not be found out or suspected. If most men or many men are capable of such plots, the sooner the earth itself explodes the better.[47]

The 1880s and 1890s saw the extensive use of dynamite among political activists too, especially Irish Fenians and anarchists in France,

Spain and England. One of the most notorious of the French anarchists was François Koeningstein, known as Ravachol, who was executed in 1892, after a versatile career as a counterfeiter, grave-robber and dynamiter. Escaping from prison, he bombed the homes of a judge and attorney involved in the sentencing of other anarchists. Ravachol seems to have become a kind of popular hero. His exploits were written up as a popular narrative,[48] and songs were sung in his honour, including one, to the tune of 'La Carmagnole', that had the refrain:

Dansons la Ravachole,
Vive le son, vive le son,
Dansons la Ravachole,
Vive le son de l'explosion![49]

Perhaps the most genuinely explosive novel relating to the dynamite decades is Joseph Conrad's *The Secret Agent*, though it did not itself appear until 1907. In a certain sense, the novel may be said to be explosive in its very form. At its heart is a grotesque accident, in which Stevie, Mr Verloc's retarded son, whom he has recruited to plant a bomb by the Greenwich Observatory, is accidentally blown up close to his target. The explosion is indeed at the novel's heart, since it is never in fact seen or described, the reader being taken immediately from the scene in which the explosion is requisitioned by the ambassador of a foreign power in order to discredit the anarchists in London, to the aftermath of the explosion that has left Stevie's mangled remains strewn over the lawns and paths of Greenwich Park.

The reader is encouraged to expect that the work of the novel will be to bring the meaning of this secret, unarticulated heart to light – to expose, explore and anatomize the stubborn clot of concealment at the heart of the novel, reintegrating what the explosion has scattered, as seems to be promised in the conversation between Chief Inspector Heat and Verloc: '"It shall all come out of my head, and hang the consequences." . . . "What's coming out?" "Everything," exclaimed the voice of Mr Verloc, and then sank very low'.[50] At times, the novel seems to suggest the world of *Bleak House*, a world that is in desperate need of the relief of explosion to clear its befuddling obfuscations. The material world is full of obstructive inertia, 'muggy stillness' (*Secret Agent*, 138), weight and mass clogging the possibility of movement and the emergence of distinction, for example in the Assistant Commissioner's view of Verloc's street:

> the van and horses, merged into one mass, seemed something alive – a square-backed black monster blocking half the street, with sudden iron-shod stampings, fierce jingles, and heavy, blowing sighs. The harshly festive, ill-omened glare of a large and prosperous public-house faced the other end of Brett Street across a wide road. This barrier of blazing lights, opposing the shadows gathered about the humble abode of Mr Verloc's domestic happiness, seemed to drive the obscurity of the street back upon itself, make it more sullen, brooding, and sinister. (*Secret Agent*, 127)

But in fact the novel shows that there is no such reintegration to be hoped for. *The Secret Agent* shows an exploded world, a world knit together by the force of its disjunction, in which emergence is in fact a merging. The novel is full of slow or stationary explosions, like the cab-ride that takes Winnie Verloc's mother to her new home:

> In the narrow streets the progress of the journey was made sensible to those within by the near fronts of the houses gliding past slowly and shakily, with a great rattle and jingling of glass, as if about to collapse behind the cab; and the infirm horse, with the harness hung over his sharp backbone flapping very loose about his thighs, appeared to be dancing mincingly on his toes with infinite patience. Later on, in the wider space of Whitehall, all visual evidences of motion became imperceptible. The rattle and jingle of glass went on indefinitely in front of the long Treasury building – and time itself seemed to stand still . . . The cab rattled, jingled, jolted; in fact, the last was quite extraordinary. By its disproportionate violence and magnitude it obliterated every sensation of onward movement; and the effect was of being shaken in a stationary apparatus like a medieval device for the punishment of crime,

> or some very new-fangled invention for the cure of a sluggish liver. (*Secret Agent*, 131, 136)

The effects of explosion are suggested throughout; for example, the man approached for charity by Mrs Verloc's mother who is 'struck all of a heap' (*Secret Agent*, 134). The novel struggles to get itself on the side of the kind of decisive action that will make distinctions and consequences plain. But it is slowly taken over by the condition of stalemate that it describes, the stalemate epitomised by the Professor's unexploded bomb, which will ensure that if he is ever taken, he will take his apprehenders with him. Explosion itself has been exploded, until it becomes the opposite of itself, an inert solidarity of dissipation. It is imaged perfectly by the London night into which the Assistant Commissioner ventures, which is described as 'an immensity of greasy slime and damp plaster interspersed with lamps, and enveloped, oppressed, penetrated, choked, and suffocated by the blackness of a wet London night, which is composed of soot and drops of water' (*Secret Agent*, 126). The oppressive atmosphere is formed from the clustering of atomized matter. The Professor himself sees the only blockage to his plans for purgative extermination of the weak as lying in the inert indifference of the mass: 'The thought of a mankind as numerous as the sands of the seashore, as indestructible, as difficult to handle, oppressed him. The sound of exploding bombs was lost in the immensity of passive grains without an echo' (*Secret Agent*, 245). We may perhaps associate the dull, generalised, subliminal fulmination of the explosion that has already happened with the all-assimilating 'boum' of Forster's Marabar Caves.[51]

This is very far from the rapturous welcome given to an explosive sensibility by Walter Benjamin:

> Our taverns and our metropolitan streets, our offices and furnished rooms, our railroad stations and our factories appeared to have us locked up hopelessly. Then came the film and burst this prison-world asunder by the dynamite of the tenth of a second, so that now, in the midst of its far-flung ruins and debris, we calmly and adventurously go traveling.[52]

Benjamin is referring here to the effects obtainable via optical tricks and techniques, especially those of close-up and its temporal equivalent, slow-motion, which enable us to see what had previously not been disclosed to human vision. An exploded view in anatomy or architecture gives a view of the different levels of a structure separated and distanced from each other in order that each may be studied in isolation. Explosion here means disembedding, disintrication, unfolding. The idea that the world of appearance is itself just such an unfolding of what had previously been mingled, or integrally knit, is older than one might think. The 'implicate ordering' of David Bohm's cosmology recalls details of the extraordinary doctrine of encapsulated preformation that held sway during the seventeenth and eighteenth centuries, which held that every living form encloses within itself, not only *in potentia*, but also in miniature, the embryonic forms of its future offspring, just as it was itself similarly contained in the womb of the first progenitor.[53]

There is a strong association between explosion and the exposure to view of different bodies and processes. Explosion allows the eye access to and jurisdiction over processes that are ordinarily too hidden or compacted to be seen. This kind of vision is not so much invasive or insinuating as enlarging or projecting: it constructs the inner space – of the body, the network, the building – as an outward space, one that it is possible to inhabit, explore.

Though one or two writers have dwelt on the 'explosion of space' that Benjamin points to here, his striking metaphor has remained more or less unremarked. But, as it happens, there is a more than metaphorical relationship between explosion and the optical unconscious of film. One of the very first applications of motion photography, even before the development of cinema, was to the photographing of explosions. An experimenter named Abbott took a series of photographs of the impact caused by submarine torpedoes. The device he used to obtain his photographic sequence was a keyboard, in which the first key triggered the explosion, and subsequent keys opened the shutter for six rapid photographic exposures. Playing the explosive arpeggio allowed the explosion to be itself photographically exploded:

> The first experiment was with the explosion of 500 pounds of dynamite, estimated equal to 5,000 pounds of gunpowder, and the pictures taken at intervals of one 10th of a second, so that all the successive pictures were taken in not much more than half a second. This is not even a very rapid succession, as almost any pianist can easily play twice as many successive keys in that time. The result was an exposition in the pictures of all the successive results, analyzed and in order . . . The eye saw nothing but a confused outburst of water, by reason of the persistence of images on the retina; but the photographic camera was very much quicker than the eye, as proved by the series of photographs, which showed the whole manner in which the hull yielded to the shock, the shape and position of the different fragments while flying up in the air and coming down again.[54]

Once cinema came along, it maintained this elective affinity with the explosion. One of the earliest films produced by Cecil M. Hepworth was *Explosion of a Motor Car* (1900), which shows a car approaching the camera along a country road. As it gets close, it explodes. A policeman rushing to the scene inspects the sky with a telescope, until shattered body parts begin to rain down around him. He gathers them together, writes a report and leaves. Hepworth would go on, reasonably enough, to make *How It Feels to Be Run Over* later that same year.

One might posit a typology of cinema that runs along a spectrum of touches or contacts. At one end, there would be the kiss: the line would extend through the slap and the punch to the explosion. The explosion has been given a new lease of life by the Internet. One of the most popular search categories on YouTube is 'explosion'. Users feverishly deposit and withdraw footage of fireballs in the Nevada desert, fireworks factories going up, and home-made combustion devices of all kinds.

'Les Empéreurs proposent; la dynamite décompose' boomed an anarchist pamphlet of the 1880s.[55] It is the force of deformation and decomposition that makes the explosive blast such an alluring subject and analogue for modernist art. Wyndham Lewis's magazine *Blast!* shares a title with a San Francisco anarchist newspaper that began publication in 1916. And yet the explosion is more than just the violent dissolution of form. From the First World War onwards, artists became captivated by the task of representing the explosion. In the cinema, which seemed to have as its secret vocation the desire to make explosions visible, legible, iterable and, perhaps most importantly of all, reversible, explosions are not only rendered, but also entered into. (This is literally the case with cinema, since early celluloid film is itself so chemically unstable, and liable to explode spontaneously.) Explosive art enables us to see explosion as a form, an informing process. This is perhaps the greatest scandal of the atomic bomb, that the images of its explosion should be so irresistibly stately and magnificent.

Explosions are caused by the sudden expansion into space of matter under pressure. But it is as a perturbation of time that they have functioned in art, and especially in the art of cinema. The explosion not only compresses matter, it also concentrates and accelerates time, which it then releases. We can say that the explosion temporalizes matter, even as it forces matter into temporal form. Famously, Sam Goldwyn thought that a film should start with an earthquake and build to a climax. Many films, indeed, cinema itself, started with the equivalent form, of the explosion. Perhaps the principal reason why explosions have entered and articulated the modern imagination is that they seem so powerfully inaugurative. Explosions begin as well as end things. We have seen that, long before the idea of the big bang, nature had used the explosion as a means for scattering seed, for the leap into improbability, the looping in of chance to purpose. And yet an explosion can never quite be an inauguration, for an explosion is a restoration of equilibrium, a falling-due of an energetic account, as a pent-up pressure is suddenly released. An explosion is a local acceleration of the tendency of everything to come to pieces, to lose its coherence and persistence, to relapse into undifferentiated fragments. But it does it by a kind of concentration, an increscence of force and of time that is itself the very opposite of diffusion: the explosion is therefore a punctual moment of diffusion. An explosion becomes a scansion of time, establishing priority and posterity.

A time of explosion is a catastrophic time, one that is always out of joint, in that it is either moving rapidly down the energy slope – in exothermic explosions – or up the slope – in endothermic explosions. But it is never in synchronicity with itself. Explosiveness – whether in the gathering of magma beneath the volcano, or the straining of tectonic plates one against another – is always ahead of itself, in that it is counting down to the next explosion, and looking forward to that discharge that will restore things to hoped-for equilibrium.

We have come to inhabit an explosive sensibility, in which we are intimately familiar with the catastrophic cadences of eruption. The spasmodic pathologies of nature have been taken across into culture, which runs to its own convulsive rhythms of boom and bust, and exponential expansion of all kinds (the population explosion, the information explosion, the various 'big bangs' of deregulation, and so on). Even the history of explosives seems to exhibit an explosive rhythm. The editor of a recent collection of essays on explosives laments that 'until the later decades of the twentieth century, explosives constituted an invisible factor in the historical process', but the foreword to the volume by the distinguished military historian and explosionist Bert Hall elatedly reports that '[t]o sum it up in a single sentence, gunpowder studies are expanding conceptually even more rapidly than the field's scope is widening geographically and chronologically'.[56] But these rhythms have been analysed, channelled and coordinated, in a process typified by the internal combustion engine, which smoothes and tunes multiple explosions into a docile hum, turning catastrophe to use, rotating impact into trajectory.

In a sense we may have developed an addiction to the general effects of the 'dynamite of the tenth of a second' – with what we might call the 'explosure' of universal visibility, communicability, analysability. Our world has been blown, not so much to bits, as to terabytes. Perhaps what continues to beguile is the prospect, amid all this widening gyre of exploded explosion, of some irreversible change, the once and for all, no-going-back definitive discharge or inauguration.

When a building is toppled, it retains much of its pathos and majesty, which explains some of the embarrassment and sense of uncanny power provoked by an Ozymandias, or grounded Colossus – think of the statues of Lenin that were hauled down across Eastern Europe after 1989, and then put into uneasy retirement in car-parks, or the statue of Saddam, that it was not enough to pull down, but which had to be assaulted and dismembered with desperately redoubled fury. The monument and the ruin are not opposites, but complements.

But when a building explodes, it is subject to a defeat, not by the ground, but by the air. A building that turns instantly to a cloud of dust and smoke has been swallowed up by the turbid expansiveness of its inner space. In place of the former coherence and continuity of line, there are only tiny fragments. When a building is turned to dust, blown to smithereens, the walls and the air sealed within them violently collide and collapse, to create a compound, particulate air full of dust, a dust full of air. The dynamiting of chimney stacks can produce this effect, but it was exhibited much more spectacularly by the collapse of the two towers of the World Trade Center after the attacks of 11 September 2001. Rather than falling epically headlong, like a Goliath or a Giant's Causeway, the towers furled sickeningly into themselves, like a sleeve rolling down. One witness in fact described the fall of the south tower as a grotesque act of exhalation:

> As we stood in utter disbelief the building seemed to huff, as if inhaling. It appeared to expand from what sounded and looked like unburned jet fuel exploding, then exhaled and collapsed straight to the ground in a slow motion mushroom cloud of dust and debris that slipped between the adjacent buildings and out into the river in front of us.[57]

The towers did not crash to the ground, but rather slid into supersaturated lightness, turning, as they went, into smoke and dust. The sky remained lucidly blue, as the streets were swallowed in a thick, swilling cumulonimbus of dust. The site was quickly named 'ground zero', but what had occurred was not really grounded at all. Rather it was an airburst, a burst in the air, and the bursting out of air.

Everything about the 11 September attacks seemed to speak of this strange dissolution or supersession of the ground. A building was attacked by an aircraft, which seemed in its turn to make of the target a thing of air. Many of those who died preferred to leap from the building than to die gulping the searing air in the crumbling tower.

The condition of their deaths and the unlocatablity of their remains, evoke other appalling vaporizations: the firestorms of Dresden, the mushroom cloud of Hiroshima, and the industrial disposal into the air of the bodies of murdered Jews, as grimly transformed in Paul Celan's poem 'Todesfuge'.

The big bang theory of the origin of the universe provides the most encompassing confirmation of this image of form and complexity, not as dissipated by explosion, but as its very form and outcome. Recent measurements of the furthest galaxies seem to show that they are not only receding from us, but also actually accelerating in their recession – the further from us, the greater the acceleration. Perhaps the strangest thing about this theory, the exact opposite of the theory of coagulating nebulae which held good for much of the nineteenth century, is its new amenability. We are not the fall-out, aftermath or residue of the big bang; we are in its midst. We are its unfolding, epileptically taken over by the form that explosion takes.

ten

The Fizziness Business

A Mouthful of Sweet Air

You can't live on air, we sometimes hear, though perhaps we do not always hear the echo in that phrase of Hamlet's words: 'I eat the air / Promise-crammed'.[1] Hamlet is here comparing himself to the chameleon, which in the early sixteenth century some still believed, on the authority of Pliny, Tertullian and others, ate nothing at all, but derived all its nourishment from the air. There are many stories of saints, ascetics and other whole-hog abstentionists who have attempted to live without food. The seventeenth-century Rosicrucian John Heydon, who believed in the efficacy of odours and aromas in calling forth angels and spirits – 'for truly the spirituall body is very much incrassated by them, and made more gross: for it liveth by vapours, perfumes and the odour of sacrifices'[2] – counselled his followers that eating solid food was in fact man's original sin, and that the pious soul ought to be able to live perfectly well on the aromas rising from a dish of meat placed on the stomach.

But there is another way besides inanition of living on air, and this is my burden in what follows. Gaston Bachelard refers in his *Air and Dreams* to the belief that 'to fly we need wings less than we need winged substance or wing-like food',[3] and reminds his readers of the angel Raphael's explanation of angelic alimentation in Milton's *Paradise Lost*:

time may come when men
With angels may participate, and find

No inconvenient diet, nor too light fare:
And from these corporal nutriments perhaps
Your bodies may at last turn all to spirit,
Improved by tract of time, and winged ascend
Ethereal.[4]

I am going to consider some of the faiths and fancies that ramify from the ambition to consume levity, to cram oneself with vacancy.

Though the desire to represent heavy and 'filling' foods as light has become a conspicuous feature of the design and marketing of sweets, cakes and puddings in the last twenty years, the desire that foods should be light as well as substantial has a long history. I have briefly reviewed the history of the notion in my *The Book of Skin*, from the viewpoint of the idea of the 'delicacy', and the desire that food should be so subtle and evanescent that one might truly have one's cake and refrain from eating it.[5] But delicacy is a matter not just of taste, but of consistency, not just of refinement, but of weight, or rather, lightness. Delicacies are always in a sense magical foods and may involve the folding in of air and the lightness it connotes through artifice, whether mechanical (beating and whisking, as in pancakes, omelettes and soufflés) or chemical, as in the carbonation of fizzy drinks which began in the late eighteenth century, and which will pop up later in this essay. But perhaps the most magical foods are those that come about through natural processes of fermentation.

Leaven

It is hard to appreciate how important processes of fermentation were, and how powerful and far-reaching the metaphor of fermentation was in other historical periods. Bachelard suggests that fermentation represents the blending of airiness or spirit with the resistance of matter, and thus a spiritualizing of the 'primary paste', which is the ideal form of the imagination of matter.[6] If human beings blend their energies with matter in forms of primary paste, folding air into earth, then fermentation appears to be the process in which the earth spontaneously moulds or transforms itself. It is not energy applied to matter, but an interior energy that spills or breeds out of matter itself. It is the way in which matter 'works'.

Fermentation is closely associated with processes of transmutation, and this accounts for the centrality of bread and wine in Christianity, as well as for fermented foods and beverages in other religions, such as the soma of book Nine of the *Rig-Veda*. While not as common as discoverers of fire, myths of culture-heroes who discovered how to control the processes of fermentation are widespread. For many cultures, the first fermented alcoholic drink is likely to have been a form of mead, which is made by mixing honey with water, and allowing it to ferment for a few days. Honey itself has a kind of airy aspect. The bees who produced it were themselves long believed to have been produced through spontaneous generation (from the blood of oxen), which is a kind of parallel to the process of fermentation. In 1 Samuel 14 the battle-weary and starving Israelites come upon a wood where honey is miraculously provided: 'And all they of the land came to a wood; and there was honey upon the ground. And when the people were come into the wood, behold, the honey dropped' (1 Samuel 14: 25–6). This seems to associate honey with the manna that feeds the Israelites for forty years in their wanderings from Egypt:

> [I]n the morning the dew lay round about the host. And when the dew that lay was gone up, behold, upon the face of the wilderness there lay a small round thing, as small as the hoar frost on the ground. And when the children of Israel saw it, they said one to another, It is manna: for they wist not what it was. And Moses said unto them, This is the bread which the LORD hath given you to eat. (Exodus 16: 13–15)

Indeed the airdropped manna is said to taste 'like wafers made with honey' (Exodus 16: 31).

A bee plays a decisive part in the account of the primal brewing of beer in the Finnish *Kalevala*. Rune 20 tells how Osmotar, the woman charged with preparing the beer for a grand wedding, mixes together hops, barley and water, but is unable to make them ferment, and utters the following appeal, as rendered in the Hiawatha-rhythms of John Martin Crawford's translation:

What will bring the effervescence,
Who will add the needed factor,
That the beer may foam and sparkle,
May ferment and be delightful?[7]

Various magic creatures are produced from splinters and sent off to find the missing ingredient. A squirrel brings fir-cones, a marten gathers the angry foam from the lips of warring bears, but both leave the beer 'cold and lifeless'. Finally, a honey-bee is generated (by rubbing a pea-pod on the thighs of a young virgin, how else) and sent far overseas to gather nectar from special flowers and grasses:

Osmotar, the beer-preparer,
Placed the honey in the liquor;
Kapo mixed the beer and honey,
And the wedding-beer fermented;
Rose the live beer upward, upward,
From the bottom of the vessels,
Upward in the tubs of birch-wood,
Foaming higher, higher, higher,
Till it touched the oaken handles,
Overflowing all the caldrons;
To the ground it foamed and sparkled,
Sank away in sand and gravel.[8]

The working of the ferment not only brings the beer to life, it also incites in it a voice, one of the many dream-identifications between humans and the substances they eat. Battened down within wooden casks, the beer itself begins to sing of its desire for a singer to come and carol its praises:

Stronger grew the beer imprisoned
In the copper-banded vessels,
Locked behind the copper faucets,
Boiled, and foamed, and sang, and murmured:
'If ye do not bring a singer,

> That will sing my worth immortal,
> That will sing my praise deserving,
> I will burst these bands of copper,
> Burst the heads of all these barrels;
> Will not serve the best of heroes
> Till he sings my many virtues.'[9]

The process of fermentation is highly ambivalent, for it brings together what many cultures have tended to see as opposites: putrefaction and transmutation. Fermenting milk, fruit and sugar all produce rottenness, but also valuable new products – cheese, wine and beer. This ambivalence is concentrated in Christian attitudes towards yeast or leaven. In general, Christianity inherits from Judaism a suspicion of leaven, especially in its use in bread. As Jean Soler has observed, leaven alters the nature of the substances to which it is added, and is thus to be regarded as unclean. It is to be contrasted with salt, which far from being forbidden, is mandatory in ritual offerings: 'ye shall burn no leaven, nor any honey, in any offering of the LORD made by fire' commands Leviticus (Leviticus 2: 11). Where leaven alters (Leviticus 2: 13), salt preserves and consolidates.[10] When leaven appears in the New Testament, however, it seems to have more positive associations. Luke 13: 20–21, for example, reads: 'Whereunto shall I liken the Kingdom of God? It is like leaven, which a woman took, and hid in three measures of meal, until the whole was leavened.' In a sermon giving a history of the church in England, W. E. Scudamore elaborated the parable, seeing in it an intention to figure 'the effectual working of the Spirit of grace upon a heart submissive to His power'. A point for point explication of the analogy between fermentation and the effects of grace follows:

> The leaven is hidden in the meal, but its presence is soon known by its effects. There is a movement as of life within the fermenting mass: it swells and heaves, and, as if endowed with life indeed, takes to itself the properties of that which acts upon it. Thus is it with an awakened soul under the secret operation of Divine grace. A quickening influence pervades it, subduing all things in it to itself. It is agitated by new desires: it hopes and trembles, and finds no

> repose until the holy, saving change is wrought, and through grace it hath become gracious.[11]

Others used etymological evidence to support the positive associations of yeast. E. W. Bullinger alleged a link between yeast and German *Geist* and related words like *ghost* and *gist*, as well as words denoting airy substance or action, like *gas* and *gust*, because of a reference he finds 'to the working of some invisible power, like the "power of the air", exciting internal motion, and producing the effect of *foaming* or *frothing*'. He even suggests a link between the words *yeast* and *Easter* on account of the shared associations with rising.[12] Alfred Austin used the metaphor of leaven in a similar way, to suggest the uplifting and enlarging effects of religious faith:

> Man needs some leaven for his daily life,
> That else were sad to heaviness, some barm
> By whose fermenting may his fancy rise
> Beyond the level of confining fact;
> And for the lightening of simple souls
> There's no such yeast as faith.[13]

However, others made out a different meaning for leaven. Bullinger noted that 'leaven' appears to have a positive meaning only in three of the biblical uses of the word: Leviticus 23: 17, Amos 4: 4–5 and Matthew 13: 23. In other places, for example Galatians 9 and 1 Corinthians 6–8, '[t]he work of this leaven is only evil.'[14] A more surprising reading of the term was brought forward by Alfred Jenour in an essay of 1855 entitled *The Parable of the Leaven Explained.* Jenour quarrels with the general interpretation of the parable as intimating the general propagation of Christianity throughout the world, making it clear that it is 'a substance tending to *putrefaction* and *corruption*'.[15] The term is 'invariably used as the symbol of moral depravity. It represents a corrupt principle which has the power of diffusing itself, and corrupting that which comes within its influence'.[16]

However, it is in the alchemical tradition that the idea of fermentation has been most fully developed. Alchemy is sacred materialism:

although there were religious forms of alchemy, the practice sought to find in chemistry a substitute field of operations for the spiritual. For the alchemical tradition, there was no wholly unspiritualized matter – nor, for that matter, was there a wholly immaterial notion of spirit. All matter contained within it qualities of the gross or earthly, and qualities of the airy. Many alchemists assumed the existence of a general principle equivalent to the Stoic conception of the *pneuma*, the informing fire or spirit that runs through the cosmos and, in different degrees of commixture with the colder, grosser elements of earth, accounts for the tenor of all things. We cannot overstate the importance of the notion of *composition* or *texture* in this conception of the world: what matters most fundamentally, most formatively, are the relations between the dense and rarefied. For this mode of thought, energy is not a different thing from matter but a rarefied from of it.

Fermentation was the name given to any vigorous chemical reaction, especially one with visible effects of boiling or sizzling (ferment derives from *fevere*, to boil). Fermentation was thus the general name given to the process whereby material things underwent change. As the anonymous *Philosophical Enquiry Into Some of the Most Considerable Phenomena's* [*sic*] *of Nature* (1715) put it:

> Fermentation, is that Operation of Nature, which takes place in all natural Productions, in Nutrition, Multiplication, Translation and Reduction of Every Species, and likewise, to improve, and heighten their proper Qualities, and multiply their respective Virtues. This is perform'd by the vital Spirit of the Universe (diffus'd through every Part and Member of the same) by means of an innate Ferment that is proper to every differing Member, of every differing Species, which God in his Projection and Creation of the Deep placed in the very Foundation of every seminal Virtue.[17]

Fermentation was thought to have two seemingly contrary effects. Fermentation dissolved things, breaking down their distinctness: this was often called 'putrefaction'. But fermentation could also precipitate solid forms out of liquid forms, in which case it was known as 'digestion', and thought of not as decomposition but as an in-forming. It is because of this that the separation out of bodies and forms can be ascribed to fermentation. After a long explanation of the operations of fermentation in wine or beer, the 1715 *Philosophical Enquiry* explained that

> Nature, which knows no such Bounds, does not stop here; for if she finds, either, any of the Wine, Spirit, Vinegar, Lees, or even the Rape, unoccupy'd for Humane Use, she'll presently convert it to some Use or other, for her Purposes . . .
>
> If she finds any Vinegar, she converts the Salt into innumerable fluctuating Animals.
>
> If she finds any of the Rape, she converts that to an infinite Number of Gnats, and other flying Creatures.
>
> If she finds any of the Lees, those she converts (with help of a little Heat) to creeping Creatures, like Bots, &c.[18]

But there was a higher kind of informing that ferments could bring about, in which solid objects, or living beings, could be refined into their own spiritual essences. The most important and philosophically significant of the many misunderstandings of chemical processes in alchemy was the much-repeated doctrine that fermentation released energies or potentials that lie latent in every form of matter. 'Ferments', which sometimes mean catalysts, rather than chemical reagents, were believed to bring about their effects in this way. Alchemists often found spiritual analogies in the operations of brewing. The Benedictine monk Basil Valentine explained the operations of yeast on beer in his *Triumphant Chariot of Antimony* of 1604:

> . . . a little *Yest* or *Ferment* is added, which excites an internal motion and Heat in the Beer, so that it is elevated in it self, and (by the help of time) *Separation* of the dense from the rare, and of the pure from the impure is made; and by this means the Beer acquires a constant virtue in Operating, so that it penetrates and effects all those *Ends*, for which it was made and brought into use: which before could not have been; because the Spirit,

> the Operator was hindred, by its own Impurity, from effecting its proper Work.[19]

Fermentation is thus regarded as the paradoxical process whereby a foreign agent, brought into contact with a body, releases or develops in that body its own innate or instinctual spiritual principle. John Heydon declared that '[f]ermentation is when anything is resolved into itself, and is rarified, and ripened: whether it be done by any ferment added to it, or by digestion only'.[20]

The visible and audible effects of fermentation – warming, hissing, foaming, formation of bubbles and, most of all the effect of increase in volume, the magical developing of a substance apparently from within itself – seemed to make the assumption irresistible that some inner principle was in course of production. Thus, fermentation, which comes about through the compounding of things, was routinely misdescribed as effecting purifications and simplifications of matter. George Ripley, one of the most notable English alchemists of the late fifteenth century, wrote in his verse treatise *The Compound of Alchemy* of the operations of fermentation:

> For Fermentation true as I thee tell,
> Is of the soul with the bodies incorporation
> Restoring to it the kindly smell,
> With tast and colour by naturall conspissation,
> Of things disseuered, a due reintegration,
> Whereby the bodie of the spirit taketh impression.
> That either the other may help to haue ingression.[21]

Concoction

It was well known that fermentation of bread or alcoholic drinks could be retarded by keeping them out of the air. Exposed to the air, by contrast, the decomposing matter was quickened by invisible agencies, which were sometimes thought of as airborne ferments (insofar as this doctrine anticipates the discovery of the microscopic spores responsible for fermentation, it is prescient). This connects with the surprisingly long-lived belief in the airy nature of impregnation. The wind was believed to have the power literally to inseminate certain creatures, such as mares and vultures. But, even where the air or wind was not literally responsible, the life-giving principle was held to be airy. Aristotle maintained in his *Generation of Animals* that sperm was infused with *pneuma*, which both embodied the vital heat of the universe and the formative impulse that enables differentiation, and the movement of the potential into the actual.[22] Aristotle closely associated sexual pleasure with the production of this generative principle: 'The pleasure which accompanies copulation is due to the fact that not only semen but also *pneuma* is emitted.'[23] This caused a certain perplexity that lasted until the seventeenth century. For if, as seemed plainly apparent, women also experienced pleasure from the sexual act, and also produced emissions, should this not indicate that they too produce a generative or informing principle? Aristotle judges not. As so often, it comes down to eating and a certain kind of indigestion: 'the female, in fact, is female on account of inability of a sort, viz., it lacks the power to concoct semen out of the final state of the nourishment (this is either blood, or its counterpart in bloodless animals) because of the coldness of its nature'.[24]

It is the notion of cooking that brings together these material processes in the world and the human body. Well into the eighteenth century the process of ingestion was understood in terms that chime with alchemical and other understandings of (what we would now call) chemical change. This understanding of the nutritive process was almost an opposite to that which prevails now. Rather than food being understood as building or sustaining the physical frame of the body, or as providing fuel for its exertions, food was understood primarily in terms of the process of rarefaction. Digestion was thought of as a series of heatings or cookings, which separated the purer, or airier constituents of food from its grosser. Theories of digestion superimposed upon the humoral theory, with its permutable quaternary of hot and cold, dry and moist, a simpler scheme of the gross and the refined, or the solid and the airy. Those foods that were most health-giving were those that encouraged the passage of the gross into the ethereal, the progressive refinement of food, into chyle, then humours, then the three orders of spirits: natural, vital and animal. Approved foods tended to be those that are easy to digest and encourage attenuation. Dubious or disparaged foods tended to be those that caused sluggishness or putrefaction (which was usually identified with stoppage or obstruction). The concern with venting, preventing or loosening various kinds of clotting or congestion was intense in the medieval and early modern periods, and is never far away in the magical conceptions of the body that still prevail in the modern world.

The conviction that the process of digestion involved a kind of rarefaction encouraged a discourse of abstinence. It was important not to clog or burden the body. Indigestion was believed to be not just an inconvenience, but a strain on and danger to the organism, since it distorted or threatened a fundamental principle of its workings. Our assumption is that earlier periods did not have to deal with the problems of surfeit that have been produced by the production of economic surplus in capitalism. However, most diet books of the late medieval and early modern period stress, not so much the dangers of malnourishment, as the dangers of gluttony and over-indulgence, along with the benefits to health of temperance and abstinence:

> almost all the diseases, with which men are ordinarily vexed, have their beginning and birth from Repletion; that is to say, from mens taking more of meat and drink, then Nature requires, and then the stomack can perfectly concoct. In proof whereof we see, that almost all diseases are cured by Evacuation. For bloud is taken away either by opening a vein, or by cupping glasses, leaches, or otherwise, that Nature may be lightened.[25]

Michael Scot's *Mensa philosophica* answered the question 'What is the Physicians best rule for health?' with 'Temperance, avoyding satiety and fulnesse', and reminded its readers of the legendary King Cyrus who 'never sate downe without a stomacke, nor never rose without an emptinesse'.[26] Dullness, heaviness, and lethargy were regarded as taking a dangerous toll upon the energies of the organism: 'If so be thou take so much meat and drink, as thou afterwards findest a certain kinde of dulnesse, heavinesse, and slothfull wearinesse, whereas before thou wast quick and lightsome; it is a signe, that thou hast exceeded the fitting measure.'[27]

In general, foods that conduced to the production of spirit, rather than the stubborn residues of crudities, or undigested food, were preferred. However, there were good and bad kinds of endogenous air: windiness or flatulence was identified with stoppage rather than attenuation. Foods that opened the bladder, like capers, unstopped the liver, or that thinned bile, like rue, or loosened phlegm were approved. Thus, anise was thought to be good both against flatulence and to prevent vapours rising to the head.[28] There was as much concern with flatulence as there was with indigestion, and the widespread disdain, inherited from Greek medical authorities, for fruit and vegetables probably has much to do with their flatulent tendencies. Galen was reputed to have lived to be more than 100 on account of never touching fruit. Good air is air that is on its way to refinement into animal spirits. Time and again, one reads of the dangerous or uncomfortable effect of the heavy, gross or corrupted, which must be dissipated, expelled or dissolved.

For this reason, aromas were an important part of dietary guidance: the smell of roses, musk, cubebs and camomile were thought to clear the head, as were sage, mustard and pepper. Some advised sleeping with the windows shut, because the night air was cold, damp and unhealthy.[29] *The Englishmans Docter* (1607) advised that '*Mazed Braines*' were the result of 'want of vent behind'.[30] Nightmares and madness were regarded as the effect of corrupted fumes rising from imperfectly digested food, or choler which, instead of being refined into animal spirits, had been overcooked and singed black. (That fermentation is associated with madness too is suggested by the word 'barmy', which derives from 'barm', a word for the foam on the top of beer or other fermenting liquid.) Andrew Boord's influential *Breviary of Health* offered the following counsel regarding 'inflation of the brain':

> The cause is whan the poores be opened oute or aboue al natural courses, it doth let in subtyll wynde ye which doth make inflacion, or else the pores opened, coldnes descendynge from the brayne is

> reuerberated unto the ventricles of the brayne agayne & maketh inflacion which is a peryculus passion, and doth put a man in parel and ieopardy of death.[31]

Corrupting and contagious airs were widely feared, while sweet scents and good air were fostered and sought out. South winds were regarded as unhealthy, north and east winds as most temperate. One was advised not to go out in either cloudy or windy weather; for, in the former, the air is too stagnant, while in the latter it is too disturbed.[32]

Mass

Baked foods, bread, cakes and pastries seem to have a particularly close relationship to the economies of air and lightness. Such foods not only conduce to, but themselves also seem to constitute, a kind of phantasmal body, compact of earth and air. Bread extends over the whole spectrum of food from everyday to sacred. Transubstantiation enacts the passage between profane or daily bread and the redeemed or transfigured bread of the Mass. Bread is divided and broken, like the tortured body of Christ, and also shared among companions, whose name means those with whom one takes bread. A trencherman is one with whom one shares bread in an even more literal way, the 'trencher' being the large, flat slice of bread that served poorer medieval diners as a plate, and which would often be shared between two mess mates. Despite the fact that animal flesh is biologically so much closer to the fabric of the human body, it is bread that seems most to be regarded as a kind of quasi-human substance, not least, but also not only, in the Christian Eucharist. Bread, biscuits and cakes are often shaped into human form (corn dollies, gingerbread men). Theories of human generation tended to look to breadmaking techniques of kneading, rolling and shaping for analogies to the process of forming the human organism, in which the sperm was thought to act, not just as a kind of vitalizing ferment, but also as a plastic or fixative power, imposed upon the otherwise chaotic and insensate female matter. Some of this anthropomorphism is applied also to grain that is fermented into alcoholic drink. The English folksong 'John

Barleycorn' presents a Dionysian story of suffering and rebirth applied to an anthropomorphized barley, which comes to maturity in the field by growing 'a long, long beard', is cut down 'at the knee' by scythes, pricked 'to the heart' by sharp pitchforks, bound, ground and then brewed into liquor. Cakes often have important parts to play in rituals of transition, such as weddings and funerals, in which the social body is dissevered and reconstituted, and they often feature in fairy-tales, all the way through to the little cake marked 'Eat Me' encountered by Alice at the beginning of *Alice in Wonderland*, which brings about bodily transformation.

The distinction between sacred and profane bread is reduplicated within the category of the profane itself, in the distinction between the coarser wholemeal, brown or black breads, and the refined white breads that grew increasingly popular from the late medieval period onwards. Refinement was signalled by the whitening of bread – achieved not only by sieving or bolting and milling of the tough outer husks of the grain, but also sometimes through adulteration by bleaching agents such as white lead. Whitening was usually thought of as lightening into the bargain. Until the nineteenth century most breads relied upon the biochemical action of what Pasteur revealed to be the micro-organisms in yeast to produce the carbon dioxide gas that caused the bread to rise. In 1846 a flour was developed consisting of a mixture of a carbonate and an acid salt that produced carbon dioxide with the addition of water as a purely chemical process – the first 'self-raising flour'. Later in the century a certain Dr Dauglish invented a process that made it possible to inject dough with carbon dioxide by mixing it in sealed chambers containing the gas under pressure – a technique similar to those that had been in use for a century or so for producing 'aerated waters'.[33] This became known as 'aerated bread', which gave its name to a chain of cheap restaurants in the 1920s, the 'ABCs' or Aerated Bread Company. T. S. Eliot uses the name for a sardonic evocation of the flaccid and unconsecrated secularity of modern life at the end of his 'A Cooking Egg', where 'weeping, weeping multitudes / Droop in a hundred ABCs'.[34] Agreeing with T. E. Hulme that modern life shared with Romanticism a spurious gaseousness, Eliot was an early analyst of the horrors of modern lightness of being: 'We are the hollow men / We are the stuffed men / Leaning together / Headpiece filled with straw. Alas!'[35]

Tonic

It was not until the end of the eighteenth century that methods began to be developed for impregnating water with bubbles. One method involved producing gas from a mixture of 'pounded chalk or marble' and 'a strong vitriolic acid', the gas produced then being bubbled into water.[36] Carbonated water, with bubbles of carbon dioxide, was originally marketed as medicinal. It had been known since the seventeenth century that the bark of the Cinchona tree helped relieve the fever of malaria. It was only after the active alkaloid quinine had been isolated in 1820 that it was possible to give controlled doses of the drug. When the Schweppes company marketed their tonic water containing quinine in the 1850s, they were hoping to appeal to the Indian imperial market. In fact, quinine dissolves much better in ethanol than in water, hence the association between gin and tonic. The tonic effects of the drink clearly depended also on the associations of the effervescence. Large claims were made for carbonated drinks in the early years of their production. Aerated drinks drew on the cult of what I earlier described as the 'pneumatic sublime' which prospered at the end of the eighteenth century and encouraged belief in the curative and uplifting powers of various gases and vapours.

One of the marketing problems that early producers of aerated drinks faced was the fact that the gases with which water could be infused did not seem very appealing. Carbonic acid, the commonest agent of aeration, was well known to be poisonous. The prestige of oxygen therapy encouraged more comfortable and plausible claims. When one C. Searle patented an effervescent drink infused with nitrous oxide in 1839, he marketed it as 'Oxygenous Aerated Water'. He stressed that it was the oxygen carried in the compound that vitalised its consumer:

> [I]t exhilarates, by producing a positively increased quantity of the natural animating spirit of the system. The oxygen, or source of

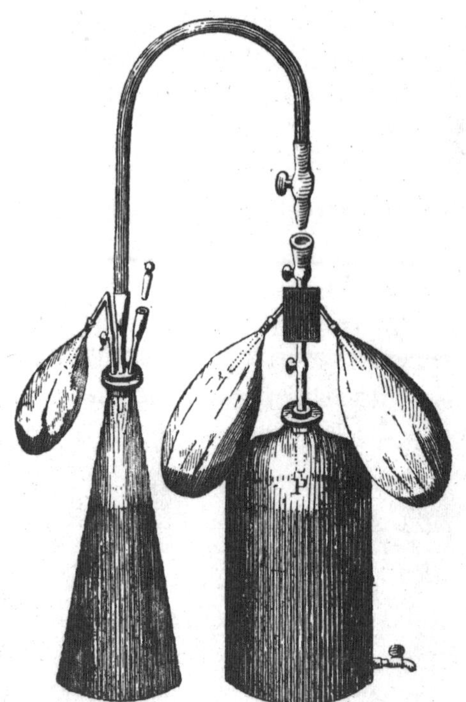

FIG. 7. Withering's Aërated Water Apparatus.

A device for producing carbonated water, from William Kirkby, *The Evolution of Artificial Mineral Waters* (Manchester, 1902).

> vital spirit, existing in this gas in the greater proportion of eight to three over its existence in the atmosphere, and being thus richer in oxygen, furnishing to those who take it a more abundant supply of animal spirits. And hence the vivid idea, the unusual disposition to muscular activity, and joyousness of feeling experienced, as noticed by Sir H. Davy and others who have tried it; as well as the cheerfulness and high spirits which succeed its use throughout the day.[37]

The water was warranted to be effective in cases of languor, debility, depression, constipation, typhus, cholera, palsy, asthma, scrofula, leprosy and scurvy, as well as 'torpor of the brain from moral dejection' and 'female obstructions'.[38] The word 'exhilarate', with which this passage begins, which dates from 1540 in English, and means literally to draw to laughter or hilarity, had by this time begun to be linked to breath and exhalation as a result of the fame of laughing gas.

I have mentioned the connection between gas and laughter. In fact, philosophy provides us with a more specific link between laughter and the history of fizz, in the only joke to be found, as far as I am aware, in the work of Kant. Here is The Joke:

> [A]n Indian at an Englishman's table in Surat, saw a bottle of ale opened, and all the beer turned into froth and flowing out. The repeated exclamations of the Indian showed his great astonishment. 'Well, what is so wonderful in that?' asked the Englishman. 'Oh, I'm not surprised myself,' said the Indian, 'at its getting out, but at how you ever managed to get it all in.'[39]

Naturally, the joke has an analytic purpose. Kant, like Freud after him, wants to show us that laughter arises from the arousing and then sudden diminishment of expectations, which itself turns out to have a gaseous dimension. We laugh, writes Kant, 'not because we think ourselves, maybe, more quick-witted than this ignorant Indian. It is rather that the bubble of our expectation was extended to the full and suddenly went off into nothing'.[40] Laughter for Kant is the exhaust formed from the differential between two nothings – a projected something, and the pricking of that expectation's bubble.

Pop

Children's stories evince a fascination with the possibilities of the bodies empowered or imperilled by lightness. Robin Kingsland's *The Fizziness Business* tells the story of two convicts, the dastardly criminal genius Oswald Bingly and his dim sidekick Stig Stubble, who escape from prison by means of a balloon, made from custard skins stitched together and inflated by a drink called 'SuperFizz'. They then steal the Crown Jewels using the custard-skin balloon to float them out of the Tower of London, but are apprehended when Stig foolishly takes a swig of SuperFizz and becomes too inflated to get through the door.[41]

Modern life has been more and more drawn to the powers of lightness, conceived not so much as the contrary of weight, as the diminution of density. Historical bodies are laden with time; they are massy, voluminous, weighed down with clothing, conformity and the habit of obedience. Modern people have made customary the experience of flight. The modern body inhabits and seeks ethereality. We seek purification through processes either of reduction or distillation. Look around any gym and you will see men attempting to turn themselves to stone and women attempting to evaporate their substance. The bodybuilder works toward petrifaction through sculpting or selective subtraction of the body (and how odd that this so solid body should be the result of what is also called 'pumping iron', as though iron itself were to be rendered in some way vaporous). The aerobic exerciser (the word 'aerobic' having been invented by Louis Pasteur in the course of his work on fermentation) seeks to turn herself to holy smoke.

Perhaps the most obvious examples of this light eating are contemporary diet foods, in which the inclusion of a deal of air is, for obvious reasons, touted as a positive good. During the 1970s the television adverts for the slimming bread Nimble famously featured a young girl who seemed not only to be desirably slender, but also so trippingly light on her feet that she seemed liable to be lifted up at any moment into the air, a condition suggested by shots of her drifting in a balloon, nibbling the product, and literalized by a freeze-frame that suspended her in mid-stride at the end of the advert. The promise was as clear as it was crazy: eat enough of this insipid stuff (the kind of thing Scousers like my dad called 'water standing up'), and you would not only not gain weight, you would also actually gain lightness, or become lighter than air, without losing any of your bodily volume. The promise was that you could become a svelte kind of human balloon, satiated by your surfeit of nothingness.

Sweetness is also often equated with lightness, for example in puff pastry, or in the candies and sweetmeats that are meant to inspire thoughts of lightness. This can be literally because they contain holes, like the Curly-Wurly bar, or Polo (the 'mint with a hole') or are full of bubbles, like Aero, or Crunchie, or because they have been whipped into supernatural afflatus (the Walnut Whip contains such an ethereal nectar poured into a cavity at the centre of the sweet). The fact that so many sweets have such inner cavities or enclosures suggests the importance of the idea of hollowness to the pleasure; Easter eggs obviously represent the most versatile kinds of improvisation upon the idea of luxurious void. Most bizarrely of all, there is a tendency for sweets to be associated with outer space: the vast interstellar vacancies that caused Pascal such disquiet cause delight when outer space is somehow thought to have entered into the inner space of the bar's composition, as in Milky Way, Starbursts, Galaxy and the Mars Bar, the last surely among the stodgiest and least ethereal confections ever made.

In the context of a discussion of the contrasting orders of the 'deep' and the 'foamy' in contemporary washing powders, Roland Barthes admirably and almost unimprovably characterizes the new function of effervescence in contemporary experience. Foam is now, not just a principle of liveliness: its most important signification is that of luxury. Foam

> appears to lack any usefulness; then, its abundant, easy, almost infinite proliferation allows one to suppose there is in the substance from which it issues a vigorous germ, a healthy and powerful essence, a great wealth of active elements in a small original volume. Finally, it gratifies in the consumer a tendency to imagine matter as something airy, with which contact is effected in a mode both light and vertical, which is sought after like that of happiness either in the gustatory category (foie gras, entremets, wines), in that of clothing (muslin, tulle), or that of soaps (film star in her bath).[42]

Barthes reminds us of the spermatic and spiritual background to the powers of foam, 'inasmuch as the spirit has the reputation of being able to make something out of nothing, a large surface of effects out of a small volume of causes'.[43] His comments make it clear that both orders, the orders of comfort and power, are economic, for both depend upon the magic of painless surplus, of something for nothing, something from nothing. His suggestions connect lightness of eating with a much larger range of experiences and images of lightness. Of course, the alchemical tradition is powerfully at work in these conceptions. But we must note that in modern conceptions the notion that governs alchemy – of a matter striving ever upwards – has been transformed. The striving of the ferment, the difficulty and the ardour of the quest, have shifted into the production of images of ease and comfort, the narcotic, nirvana bliss of being cosseted. For the alchemists, fermentation was matter energised or spiritualized; for the advertising industry, ferment is now merely Yeats's 'spume that plays / Upon a ghostly paradigm of things'.[44] Energy becomes a kind of busy, purposeless inertia. The role of the foamy in detergents is not to signal chemical or spiritual change, but to disguise it. The foamy is thus the opposite of the abrasive: 'What matters is the art of having disguised the abrasive function of the detergent under the delicious image of a substance at once deep and airy which can govern the molecular order of the material without damaging it.'[45] The point seems to be, not potentiation, but deferral and suspension. Thus, sodas, seltzers and other kinds of fizzy drinks are marketed as 'soft' drinks, with their origins in the temperance beverages of the nineteenth century. The point was to mimic the sparkle of alcoholic drinks, while offering safer and less harmful forms of exhilaration. But soft drinks have been blended with alcohol in more and more versatile ways, the alcopops of recent years being the most recent form. Once again, the fizz is in part the sign of a factitious safety, a way of mitigating the dangerous hardness or bite of alcohol.

The airy, the foamy, the frothy, the bubbly, and its allotropes, the creamy, the milky, the misty, belong to a culture of lightness and uplift that perhaps has its beginnings in the invention of comfort, signified by the new importance of the cushion in upholstery and interior design from the beginning of the eighteenth century. From this point on, furnishings will no longer frame or support the body, or augment its powers. The plumply stuffed cushions of the Baroque stagnate into the weighty upholstery of Victorian furnishing, but are newly volatilized in twentieth-century commodity culture, which markets lightness of being in every possible mode – in food, cosmetics, clothing, alcoholic drinks, perfumery and sport. Indeed, the principle of propagation that had previously inhered in the idea of ferment has become the principle of modal conflation, as fizz, foam and lightness spill across between different sensory realms and registers. One good example is the word 'pop', which, as the name for a carbonated drink, originated in nineteenth-century America, the word referring to the popping of corks. By the end of the following century, as Margaret Visser points out, the name of the drink that has come to characterize modern life, mostly through the massive spread of Coca-Cola across the world, has become wholly identified with the popular – with pop music (which, in one of its most infantile forms, is known as 'bubblegum music'), pop art and so on.[46]

Perhaps all these possibilities come together in champagne, the emblem of the modern world's embrace of lightness and frivolity. The anaesthetic qualities of champagne are embodied in the joke that Beckett tells in his *Murphy*: 'Q. Why did the barmaid champagne? A. Because the stout porter bitter.'[47] Many alcoholic drinks have natural effervescence, because of the carbon dioxide that is a by-product of the yeast's transformation of sugar into alcohol. Wine-makers had noticed that wine that had apparently completed its fermentation would occasionally be subject to a secondary fermentation in the spring following their autumn bottling, caused by spores of yeast that still lay in the imperfectly filtered wine being revived by the returning warmth and light and going to work on the residual sugar in the wine. This had sometimes been seen to occur in the wines of the region of Aÿ in France. Around 1700, the already-celebrated wine-maker Dom Pérignon succeeded in devising a method to produce and control this secondary fermentation, by adding sugar to the wine, and stacking the bottles diagonally to allow the bubbles of carbon dioxide to rise and the yeast sediment to run down to the cork, where it could be extracted little by little.[48] Champagne ('shampoo' in popular parlance, following the rule of lexical fermentation) has become a kind of everyday luxury, a vulgar refinement, or vernacular magic, the necessary accompaniment to every office party, or sporting victory. The improvident gushing or spurting of £14.99 champagne is our secularized pneumatic bliss.

The paradox of the fizziness business in the modern world is that, while we have retained much of the apparatus of thought and feeling that privileges the light body, animated by airy energy, we have turned the consumption of lightness into a kind of gourmandising. While two-thirds of the world's population inhabit literally light or malnourished bodies, the remaining third has gorged itself into obesity on the passionate dream of lightness, the effort to become, in Yeats's words 'a mouthful of sweet air'.[49]

Eating is the most conspicuous form of our bodily transactions with the world, which systematically contradict our view of the body as a self-enclosed, and enclosing, entity, and our projection of the body as a succession of uniform images or states – black white, young, aged, fit, sick, dead, etc. Eating and drinking are the primary forms of the body's traffic, not just with other bodies, but with the great, shifting, mixed body of the world, a world that, in the forms of foam, ferment and effervescence, is ever at work upon us, as well as we upon it. Eating and drinking are usually assumed to be the contrary or the complement of cognitive life and the empire of signs, and therefore to belong to the realm of mute animal need rather than cultural self-imagination. But the habitus of eating is an important part of the dreamwork whereby bodies, indeed the very notion of a body, are formed.

The dream of the effervescent or aerated body provides a way for the body to encounter, indeed to take to itself, its other: not the trivial other of other bodies, but the other of corporeality itself. The aerated body, and its dream of lightness, is a way of taking in to the body the out-of-body experience of air. The body compact with air identifies itself with the body of air – that form of matter that embodies the immaterial, the spiritual, the imaginary. It is often said that we are becoming increasingly disembodied, that our technologies, our media, our knowledges, our pleasures and our appetites, are removing us from the frail, painful, lonely, lovely exigencies of corporeal existence. But the body will neither evaporate nor be wholly discarded by the dissipative soul, since the intelligible itself is the concoction of the sensible. The dream of lightness is the recuperation, through reincorporation, of what we miscall our condition of disembodiment. We may say, as Dickens's Scrooge does when confronted by the first of his pedagogic spirits, that there is more gravy than the grave in the notion of spirit.[50] We will need a different kind of materialism, a different kind of physics to understand the nature of physical existence in our emerging conditions. But the second nature of the body formed largely of air has a longer history than we may suspect. This may help us to weigh it, to take its measure.

And this is not least because all this lightness of being can indeed become unbearable. When 'modern primitives' seek to recapture the intensity of being in the body, it is through ceremonies and ordeals that painfully and remorselessly subject the body to its own weight. Hanging from hooks embedded in flesh, or from ropes lashed around breasts, the body is forced to witness and acknowledge its own ponderability. The body is made a *memento gravitatis*. Even here, the lightness addiction lurks. For practitioners of these and other sado-masochistic arts learn that pain leads not to knowledge, but to numbness, sham pain, the feeling, at once intoxicating and anaesthetizing, of knowing that there is a pain that you no longer feel.

Absolute Levity

Eighteenth-century chemists, believing that combustion entailed the burning off of the igneous principle, phlogiston, were puzzled by the fact that objects gained rather than lost weight during burning. They sometimes resorted to the explanation that phlogiston had the property, affirmed by Aristotle in his *De Caelo*, of 'absolute levity', or negative weight: thus the burned object gained weight because it lost its lightness. By the end of the eighteenth century the doctrine of absolute levity had been abandoned, which apparently dissolved the light into the minimally heavy. In its place, however, came a concern with the achievement of an ideal levity of the body. Though imaginary, this was a secular levity. Spirit was not to be achieved in this world. For the moderns, who so often seemed, and even claimed, to have no need of such a notion, spirit was literalized in the idiom of a body that was at once fully present and entirely weightless. In previous eras, lightness had been associated with openness and expanse, while heaviness was the characteristic of dense, closed, slow bodies. The ubiquity of effervescence in the modern world has brought about a chiasmatic crossing of qualities, leading to the oxymoron of obese levity, a lightness having all the palpability and repleteness of grosser states of matter. In the Monty Python sketch, what causes the mountainous Mr Creosote to explode is the ingestion of the 'one more wafer-thin mint' pressed upon him by the Mephistophelean waiter. Thus does lightness now bear down on us.

One of the most familiar self-images of modernity is contained in the idea, not just of hovering in mid-air, but the active operation known as 'boot-strapping'. It is common to track this idea back to an episode in a German version of the Baron Munchausen story, in which the baron pulls himself up out of a swamp by his own hair, though this episode does not appear in the 1786 English version of the baron's adventures by Rudolph Raspe.[51] The earliest usage recorded by the *Oxford English Dictionary* is a reference to persons 'who had forced their way to the top from the lowest rung by the aid of their bootstraps' from Joyce's *Ulysses* in 1922,[52] but Benjamin Zimmer has brought forward a number of nineteenth-century American usages, the earliest being a mocking reference in 1834 to a design for a perpetual motion machine by the means of which its inventor, a Mr. Murphee, 'will now be enabled to hand himself over the Cumberland river or a barnyard fence by the straps of his boots'.[53] The fact that so many of the nineteenth-century usages are American or refer to Americans suggests that the primary reference of 'boot-strapping' was initially the transatlantic virtues of energetic self-fashioning. But, during the twentieth century, the term came to be extended across a wide variety of different fields, including economics, computing, biology and, following Fritjof Capra's *The Tao of Physics* (1975), cosmology, to indicate the possibility of forms emerging or transforming themselves without foundations or supplementary external input.[54]

In Romanticism, and the forms of thought it still extensively underpins, the flashing eye and floating hair of imagination were reparative antagonists to the desiccating eye of science. But for us, inhabiting a world in which so many of the most immaterial imaginings of the past have been made actual, and in which physics furnishes a fully authenticated faërie, with sylphs and sprites replaced by quarks and leptons, artists, writers and other kinds of visionary have a new vocation. As Marina Warner has suggested, in the exploration of the material forms of the soul in her *Phantasmagoria*, '[i]n a material sense, spirits are indeed channelled, and the media are here, now'.[55] But I wonder if it is quite right to see today a simple continuity or coming good of the phantasmatic traditions that Warner conjures through *Phantasmagoria*, or a simple convergence of ancient fantasy and modern physics. I am not myself certain that the immaterialization of the world leaves the function of these traditions of figuring the immaterial unchanged, and so wonder whether *Phantasmagoria* might not rather help us see beyond, or look differently at its own conclusions. '[M]odernity did not by any means put an end to the quest for spirit', Warner affirms at the beginning of her own expedition.[56] To be sure; but perhaps it has begun to give a sharply new meaning to that quest. May it not be that, where the past sought an escape from weight, we now seek a remission from lightness, or perhaps from the danger that we may, like the cartoon character who runs off a cliff, suddenly forget to remember to forget our weight, and enter the condition of 'free-fall' that world financial markets experienced in 2008? Our lightness is not a tenuous, but a paradoxically top-heavy affair. It is not really that we have become light instead of heavy; it is that we have become light as well as heavy, foamy as well as dense, our everyday avoirdupois riddled with lightness. It is not that we are nostalgic for the lost massiness of Dasein exactly, but we would be right to be perplexed at the loss of the scale in which weight and lightness could be counterposed.

The soul used to be thought of as an intimate alterity inhabiting the self. Although one's soul was one's essential being, it tended, for just that reason, to be regarded as separate, withdrawn or inaccessible. The soul was in a sense the externalized form of one's innerness, a kind of essence on elastic. One's soul is always more or other than oneself, precisely because its nature is aspiration, or movement beyond. For this reason, one's soul was not one's own; it was owed to God, or eternity, and at risk from Satanic predation. The soul is the name for the condition of not owning what you essentially are. The paradoxical lightness of the subtle body that is the soul enacts this dual condition of that which stands apart from everything we merely are. It is the bridge by means of which the self may make alliance and have transaction with the no-thing that it is. But our condition is one in which we have no choice but to be the nothingness that we are, precisely because we seem in literal fact to have melted into air, in the sense that we have become atopic, anachronic, illimitable. Under such circumstances, in which increasingly we actually *are* our indetermination, we can no longer have a relation to our souls, precisely because we are them. The air has always been used to figure and enable, not so much anthropomorphically as psychomorphically, the mind's self-relation, its effort to catch up or coincide with itself, a theme that has been explored brilliantly in Daniel Tiffany's *Toy Medium*.[57] But now, air has become much less counterfactual, much less of a beyond or an outside. The openness of the air affirmed so rapturously by Luce Irigaray has become subject to the almost universal air-conditioning of which Peter Sloterdijk writes. Air no longer figures the illimitability to which we aspire, but rather the indetermination that we are. As we enter more and more into composition with the unothered air, we find it harder and harder to have a relation with that indetermination, since one can have a relation only to that which you are not. Once, the fact that our souls were not our own allowed them to be thought of as an intimate, ownmost ineffability. Now our souls are nothing but our own, and so vanish into us and we into them; we are having to become our own self-exceeding, that we now have no choice but to 'exist', in the transitive sense proposed by Jean-Paul Sartre.[58] Thought and soul are now uncapturable and illimitable, not because they are shy, reclusive, fugitive, sprite-like, but because they are everywhere. Air, along with all its gossamer approximations, used to be the figure of thought's magical desire to capture itself; now, increasingly, air figures thought's terror of its own epidemic ubiquity. The fascination with the forms of the airy and the insubstantial that is evidenced in the conspicuous preoccupation with ethereal and nebular states and conditions – smoke, fog, haze, cloud – in contemporary art[59] may be regarded as a poignant attempt to reestablish the kind of distance from images of the extruded soul that would allow it to be thought, felt, seen and dreamed or something transcendent. The quest for spirit once meant striving to be what you were not, or to become the no-thingness that you essentially were; now that quest means striving not to have to be the unbeing that we inescapably and manifestly are. Now we may be in quest of the possibility of a quest, of a time when it was possible for the soul still to be put into question, weighed in the scale.

A 'newly-invented engine' to measure the degree of men's innate levity or gravity, from A, 'Absolute Gravity', to I, 'Absolute Levity, or Stark Fool', from John Clubbe, *Physiognomy: Being a Sketch Only of a Larger Work Upon the Same Plan: Wherein the Different Tempers, Passions, and Manners of Men, Will Be Particularly Considered* (London, 1763).

References

one: Taking to the Air

1 Luce Irigaray, *The Forgetting of Air in Martin Heidegger*, trans. Mary Beth Mader (London, 1999), p. 8.

2 Michel Serres and Bruno Latour, *Conversations on Science, Culture and Time*, trans. Roxanne Lapidus (Ann Arbor, MI, 1995), p. 121.

3 Michel Serres, *Atlas* (Paris, 1994), p. 112 [my translation].

4 Peter Sloterdijk and Bettina Funcke, 'Against Gravity: Bettina Funck Talks With Peter Sloterdijk', *Bookforum*, February/March 2005. Online at www.bookforum.com/archive/feb_05/funcke.html (accessed 12 April 2010).

5 Peter Sloterdijk, *Sphären III: Schäume, Plurale Sphärologie* (Frankfurt, 2004), p. 23 [my translation]. References hereafter to *Schäume* in text.

6 Denis Papin, *A New Digester or Engine For Softning Bones Containing the Description of Its Make and Use In These Particulars viz. Cookery, Voyages at Sea, Confectionary, Making of Drinks, Chymistry, and Dying* (London, 1681).

7 Trude Weiss Rosmarin, 'The Terms for "Air" in the Bible', *Journal of Biblical Literature*, LI/1 (1932), pp. 71–2.

8 Thorkild Jacobsen, 'Sumerian Mythology: A Review Article' [review of S. N. Kramer, *Sumerian Mythology: A Study of Spiritual and Literary Achievement in the Third Millennium BC*, Philadelphia, 1944], *Journal of Near Eastern Studies*, V (1946), pp. 151–2.

9 Aristotle, *Meteorologica*, trans. E. W. Webster (Oxford, 1923), 349a.

10 Maurice Crosland, '"Slippery Substances": Some Practical and Conceptual Problems in the Understanding of Gases in the Pre-Lavoisier Era', in *Instruments and Experimentation in the History of Chemistry*, ed Frederic L. Holmes and Trevor H. Levere (Cambridge, MA, and London, 2000), p. 90.

11 Robert Boyle, *The Works of Robert Boyle*, ed Michael Hunter and Edward B. Davis, 14 vols (London, 1999–2000), vol. III, pp. 83, 11. References hereafter to *Works* in text.

12 Bernadette Bensaude-Vincent and Isabelle Stengers, *A History of Chemistry*, trans. Deborah van Dam (Cambridge, MA, and London, 1996), pp. 22–4.

13 Charles Webster, 'New Light on the Invisible College: The Social Relations

of English Science in the Mid-Seventeenth Century', *Transactions of the Royal Historical Society*, 5th series, XXIV (1974), pp. 19–42.

14 *The Correspondence of Robert Boyle*, ed Michael Hunter, Antonio Clericuzio and Lawrence M. Principe, 6 vols (London, 2001), vol. I, p. 42.

15 *The Correspondence of Robert Boyle*, vol I, p. 46.

16 *The Correspondence of Robert Boyle*, vol I, p. 58.

17 Stephen Hales, *Vegetable Staticks: or, An Account of Some Statical Experiments on the Sap in Vegetables: Being an Essay Towards a Natural History of Vegetation. Also, a Specimen of an Attempt to Analyse the Air, by a Great Variety of Chymio-statical Experiments* (London, 1727), pp. 316–17.

18 Jeremy Taylor, 'A Sermon Preached in Christ-Church Dublin, July 16, 1663, at the Funeral of the Most Reverend Father in God, John, Late Lord Archbishop of Armagh', *Ebdomas Embolimaios: A Supplement to the Eniautos; or, Course of Sermons for the Whole Year: Being Seven Sermons Explaining the Nature of Faith and Obedience in Relation to God . . .* (London, 1663), p. 7.

19 William Oughtred, *Mathematicall Recreations; or, A Collection of Many Problemes* (London, 1653), p. 257.

20 'The Anatomy of a Monstrous PIG Communicated to the R. S. By an Ingenious Student in Physick', *Philosophical Transactions*, XIII (1683), pp. 188–9.

21 Edward Phillips, *The New World of English Words; or, A General Dictionary Containing the Interpretations of Such Hard Words As Are Derived from Other Languages . . .* (London, 1658), sig. A4v.

22 Henry Power, *Experimental Philosophy* (London, 1664), p. 57.

23 Hannah Woolley, *The Accomplish'd Lady's Delight in Preserving, Physick, Beautifying, and Cookery . . .* (London, 1675), p. 39.

24 Thomas Shadwell, *The Virtuoso: A Comedy* (London, 1676), p. 65.

25 George Thomson, *Loimotomia; or, The Pest Anatomized . . .* (London, 1666), p. 16.

26 Jean Baptiste van Helmont, *Van Helmont's Works . . .*, trans. John Chandler (London, 1664), pp. 29, 109.

27 Johann Juncker, *Conspectus Chemiae Theoretico-Practicae* (Halae Magd., 1730), p. 365.

28 *Van Helmont's Works*, p. 109.

29 George Thomson, *Ortho-methodoz Iatro-chymike; or, The Direct Method of Curing Chymically* (London, 1675), n.p. See also Paulo Alves Porto, 'O Médico George Thomson e os primeiros desenvolvimentos do conceito de gás', *Química nova*, XXIV (2001), pp. 286–92.

30 Thomson, *Ortho-methodoz*, p. 198.

31 Thomson, *Ortho-metodosz*, pp. 8–9.

32 Walter Charleton, *Two Discourses* (*Concerning the Different Wits of Men* and *The Mysterie of Vintners*) (London, 1669), pp. 144–5.

33 Samuel Parker, *A Free and Impartial Censure of the Platonick Philosophy* (Oxford, 1666). p. 63; Henry Stubbe, *A Specimen of Some Animadversions*

Upon a Book Entituled, Plus ultra or, Modern Improvements of Useful Knowledge Written by Mr Joseph Glanvill, a Member of the Royal Society (London, 1670), p. 9.

34 John Norris, *An Account of Reason and Faith: In Relation to the Mysteries of Christianity* (London, 1697), pp. 252–3.

35 John Mayow, *Medico-physical Works: Being a Translation of Tractatus quinque medico-physici*, trans. A. C. Brown and Leonard Dobbin (Edinburgh, 1907), pp. 75–7.

36 A. Rupert Hall, 'Isaac Newton and the Aerial Nitre', *Notes and Records of the Royal Society of London*, LII (1998), p. 55; Allen G. Debus, 'The Paracelsian Aerial Niter', *Isis*, LV (1964), pp. 43–61.

37 Bernadette Bensaude-Vincent and Isabelle Stengers, *A History of Chemistry*, trans. Deborah van Dam (Cambridge, MA, and London, 1996), p. 39.

38 Robert Hooke, *Lampas or, Descriptions of Some Mechanical Improvements of Lamps & Waterpoises Together With Some Other Physical and Mechanical Discoveries* (London, 1667), pp. 9–10.

39 John Howe, *The Living Temple or, A Designed Improvement of That Notion That a Good Man is the Temple of God* (London, 1675), p. 70.

40 Robert Johnson, *Enchiridion Medicum or, A Manual of Physick Being a Compendium of the Whole Art* (London, 1684), p. 110.

41 Athanasius Kircher, *The Vulcano's or, Burning and Fire-vomiting Mountains, Famous in the World, With Their Remarkables Collected for the Most Part out of Kircher's Subterraneous World* (London, 1669), p. 2.

42 Bensaude-Vincent and Stengers, *History of Chemistry*, p. 27.

43 Bensaude-Vincent and Stengers, *History of Chemistry*, p. 27.

44 Michel Serres and Hari Kunzru, 'Art, Writing: Michel Serres (Interview with Michel Serres, 10 January 1995)'. Online at www.harikunzru.com/michel-serres-interview-1995

45 Gaston Bachelard, *Air and Dreams: An Essay on the Imagination of Movement*, trans. Edith R. Farrell and C. Frederick Farrell (Dallas, 1988).

46 Roy Hattersley, 'Pakistan Can Work It Out', *Guardian* (20 August, 2007), p. 25.

two: A Very Beautiful Pneumatic Machinery

1 Vicktor Tausk, 'On the Origin of the "Influencing Machine" in Schizophrenia', in *Sexuality, War and Schizophrenia: Collected Psychoanalytic Papers of Victor Tausk*, ed. P. Roazen, trans. E. Mosbacher, et al. (New Brunswick and London, 1991), p. 187.

2 Tausk, 'On the Origin of the "Influencing Machine" in Schizophrenia', p. 208.

3 Tausk, 'On the Origin of the "Influencing Machine" in Schizophrenia', p. 210.

4 Mike Jay, *The Air Loom Gang: The Strange and True Story of James Tilly Matthews and his Visionary Madness* (London and New York, 2004).

5 John Haslam, *Illustrations of Madness*, ed. Roy Porter (London and New

York, 1988), p. 19. References hereafter to *Illustrations* in the text.
6 Daniel Paul Schreber, *Denkwuürdigkeiten eines Nervenkranken*, ed. Gerhard Busse (Giessen, 2003), p. 126.
7 Thomas Percival, 'On the Solution of Stones . . . By Water Impregnated With "Fixed Air"', in *Philosophical, Medical and Experimental Essays . . . To Which is Added An Appendix; Containing a Letter to the Author from Dr Saunders, on the Solution of Human Calculi* (London, 1776), pp. 131–57; Thomas Beddoes, *Observations on the Nature and Cure of Calculus, Sea Scurvy, Consumption, Catarrh, and Fever: Together With Conjectures Upon Several Other Subjects of Physiology and Pathology* (London, 1793), pp. 7–10.
8 Tausk, 'Influencing Machine', pp. 193–4
9 John Haslam, *Observations on Madness and Melancholy: Including Practical Remarks on Those Diseases; Together With Cases: And An Account of the Morbid Appearances on Dissection*, 2nd edn (London, 1809), p. 68. References hereafter to OMM in text.
10 Haslam, *Observations on Madness and Melancholy*, p. 68.
11 John Haslam, *Observations on Insanity: With Practical Remarks on the Disease, and an Account of the Morbid Appearances on Dissection* (London, 1798), pp. 74, 82, 97.
12 Thomas Beddoes, *Hygëia; or, Essays Moral and Medical, on the Causes Affecting the Personal State of Our Middling and Affluent Classes*, 3 vols (Bristol, 1803), vol. III, p. 13. References hereafter to *Hygëia* in text.
13 Thomas Beddoes, *A Letter to Erasmus Darwin, MD on a New Method of Treating Pulmonary Consumption, and Some Other Diseases Hitherto Found Incurable* (Bristol, 1793), p. 9.
14 Dorothy A. Stansfield, *Thomas Beddoes MD 1760–1808: Chemist, Physician, Democrat* (Dordrecht, Boston, MA, and Lancaster, 1984), p. 218.
15 Beddoes, *Letter to Erasmus Darwin*, pp. 20–21.
16 Neil Vickers, *Coleridge and the Doctors, 1795–1806* (Oxford, 2004), pp. 46–60.
17 Beddoes, *Letter to Erasmus Darwin*, p. 11.
18 Beddoes, *Letter to Erasmus Darwin*, pp. 11–12.
19 Beddoes, *Observations on the Nature and Cure of Calculus*, pp. 147–8.
20 Beddoes, *Observations on the Nature and Cure of Calculus*, p. 148.
21 Erasmus Darwin, *The Golden Age: A Poetical Epistle From Erasmus D---n to Thomas Beddoes* (London, 1794), p. 10.
22 Quoted in Jay, *Air Loom Gang*, p. 251.
23 Joseph Priestley, *Experiments and Observations on Different Kinds of Air* (London, 1790), vol. I, p. xxiii.
24 Quoted in Trevor H. Levere, 'Dr Thomas Beddoes and the Establishment of his Pneumatic Institution: A Tale of Three Presidents', *Notes and Records of the Royal Society of London*, XXXII (1977), p. 44.
25 Thomas Beddoes, letter to James Watt snr, quoted in Dorothy A. Stansfield and Ronald G. Stansfield, 'Dr Thomas Beddoes and James Watt: Preparatory Work, 1794–96, for the Bristol Pneumatic Institute', *Medical History*,

xxx (1986), p. 285.
26 Edmund Burke, *Reflections on the Revolution in France*, ed. Frank M. Turner (New Haven, CT, and London, 2003), p. 7.
27 Guyton de Morveau, Antoine Lavoisier, Claude Louis Berthollet and Antoine de Fourcroy, *Méthode de nomenclature chimique* (Paris, 1787).
28 Guyton de Morveau et al., *Method of Chymical Nomenclature*, trans. James St John (London, 1788), pp. 128–30.
29 Joseph Priestley, *Considerations on the Doctrine of Phlogiston and the Decomposition of Water* (Philadelphia, 1796), pp. 9, 10.
30 Matthew Dobson, *A Medical Commentary on Fixed Air*, 3rd edn (London, 1787).
31 Royall Tyler, *The Yankey in London: Being the First Part of a Series of Letters Written by an American Youth, During Nine Months Residence in the City of London* (New York, 1809), p. 135.
32 Burke, *Reflections on the Revolution in France*, p. 134.
33 Samuel Taylor Coleridge and Robert Southey, *The Fall of Robespierre*, Coleridge, *Complete Poetical Works*, ed. Ernest Hartley Coleridge, 2 vols (Oxford, 1912), vol. II, p. 502.
34 Thomas Beddoes, *Notice of Some Observations Made At the Medical Pneumatic Institution* (Bristol, 1799), p. 15.
35 Humphry Davy, *Researches, Chemical and Philosophical; Chiefly Concerning Nitrous Oxide; or, Dephlogisticated Nitrous Air, and its Respiration* (London, 1800), p. 508.
36 Davy, *Researches, Chemical and Philosophical*, p. 509.
37 Beddoes, *Notice of Some Observations*, p. 10.
38 Beddoes, *Notice of Some Observations*, pp. 16–18.
39 Beddoes, *Notice of Some Observations*, p. 17.
40 James Kendall, *Humphry Davy: 'Pilot' of Penzance* (London, 1954), p. 46.
41 Davy, *Researches, Chemical and Philosophical*, pp. 487–9.
42 Bedddoes, *Notice of Some Observations*, p. 27.
43 Richard Polwhele, 'The Pneumatic Revellers: An Eclogue', *Poems*, 5 vols (Truro, 1810), vol. V, pp. v–vi. References hereafter to 'Pneumatic Revellers' in text.
44 Vickers, *Coleridge and the Doctors*, pp. 34–5.
45 Beddoes, *Observations on the Nature and Cure of Calculus*, p. 258.
46 Roy Porter, 'Mesmerism in England', *History Today*, XXXV (1989), p. 27.
47 Armand Mari-Jacques de Chastenet, Marquis de Puységur, *Mémoires pour servir à l'histoire et à l'établissement du magnétisme animal*, ed. Georges Lapassade and Philippe Pédelahore (Toulouse, 1986), p. 9.
48 *Mesmerism: A Translation of the Original Scientific and Medical Writings of F. A. Mesmer*, trans. George Bloch (Los Altos, 1980), pp. 13–14. References hereafter to *Mesmerism* in text.
49 Daniel C. Dennett, *Darwin's Dangerous Idea: Evolution and the Meanings of Life* (London, 1995), p. 63.
50 Gilles Deleuze and Félix Guattari, *A Thousand Plateaus*, trans. Brian Massumi (London, 1999), pp. 471–500.

51 Hartmut Kraft, *Grenzgänge zwischen Kunst und Psychiatrie* (Cologne, 1986), pp. 55–6.
52 Jay, *The Air Loom Gang*, pp. 29–34.
53 Joseph Conrad, *Collected Letters*, ed. Frederick R. Karl and Laurence Davies, 9 vols (Cambridge, 1983–2007), vol. I, p. 425.
54 Friedrich Krauss, *Nothschrei eines Magnetisch-Vergifteten (1852) und Nothgedrungene Fortsetzung meines Nothschrei (1867): Selbstschilderungen eines Geisteskranken*, ed. H. Ahlenstiel and J. E. Meyer (Göttingen, 1967). References hereafter to *Nothschrei* in text. Further selections appear in *Grenzgänge zwischen Wahn und Wissen: zur Koevolution von Experiment und Paranoia, 1850–1910*, ed. Torsten Hahn, Jutta Person and Nicolas Pethes (Frankfurt and New York, 2002), pp. 35–57. References hereafter to *Grenzgänge* in text.
55 D. G. Kieser, *System des Tellurismus oder thierischen Magnetismus: ein Handbuch fur Naturforscher und Aerzte*, 2nd edn (Leipzig, 1826).
56 *Perceval's Narrative: A Patient's Account of his Psychosis, 1830–1832*, ed. Gregory Bateson (Stanford, CA, 1961), pp. 113–14. References hereafter to *Perceval* in text.
57 Jay, *Air Loom Gang*, p. 226.

three: Inebriate of Air

1 Sigmund Freud, 'The Uncanny' [1919], trans. Alix Strachey, in *Pelican Freud Library*, vol. 14: *Art and Literature*, ed. Albert Dickson (London, 1988), pp. 362–3.
2 Sigmund Freud, *Totem and Taboo* [1913], trans. James Strachey, in *Pelican Freud Library*, vol. 13: *The Origins of Religion*. ed. Albert Dickson (London, 1985), pp. 143–4.
3 René A. Spitz, 'The Genesis of Magical and Transcendent Cults', trans. Hella Freud Bernays, *American Imago*, XXIX (1972), p. 2.
4 Freud, *Totem and Taboo*, p. 146.
5 Theodor Reik, 'On the Effect of Unconscious Death Wishes', trans. Harry Zohn, *Psychoanalytic Review*, LXV (1978), pp. 38–67.
6 Michel Serres, *Variations sur le corps* (Paris, 1999), p. 12 (my translation).
7 Emily Dickinson, *Complete Poems*, ed. Thomas H. Johnson (London, 1984), p. 99.
8 Samuel Sexton, 'The Preservation of Hearing', *Harper's New Monthly Magazine*, LX (1879–80), p. 615.
9 Annie Paysan Call, *As A Matter of Course* (London, 1895), pp, 20–21.
10 John Marshall, *On A Circular System of Hospital Wards* (London, 1878), quoted in Jeremy Taylor, 'Circular Hospital Wards: Professor John Marshall's Concept and its Exploration By the Architectural Profession in the 1880s', *Medical History*, XXXII (1988), p. 427.
11 Thomas Robinson, *Breathing: Considered in Relation to the Bodily, Mental and Social Life of Man* (Glasgow, 1869), p. 23.
12 T. S. Eliot, *Complete Poems and Plays* (London, 1969), pp. 37–8.

13 Arthur H. Ewing, *The Hindu Conception of the Functions of the Breath. A Dissertation* (Baltimore, 1901).
14 Leontes says 'lawful as eating': William Shakespeare, *The Winter's Tale*, ed. J.H.P. Pafford (London and Cambridge, MA, 1963), vol. III, 110–11, p. 159.
15 W.D.A. Smith, *Under the Influence: A History of Nitrous Oxide and Oxygen Anaesthesia* (London and Basingstoke, 1982), pp. 34–40.
16 James John Gareth Wilkinson, *Epidemic Man and His Visitations* (London, 1893), pp. 77, 78.
17 Humphry Davy, *Researches, Chemical and Philosophical; Chiefly Concerning Nitrous Oxide; or, Dephlogisticated Nitrous Air, and Its Respiration* (London, 1800), p. 489.
18 William Hamilton Drummond, *The Pleasures of Benevolence: A Poem* (London, 1835), pp. 27–8.
19 William Ramsay, 'Experiments with Anaesthetics', *Journal of the Society for Psychical Research*, VI (1893), p. 94.
20 William James, 'On Some Hegelisms', *Mind*, VII (1882), p. 206. References hereafter to 'On Some Hegelisms' in text.
21 William James, 'Recent Literature', *Atlantic Monthly*, XXXIV (1874), p. 628.
22 James, 'Recent Literature'.
23 Benjamin Paul Blood, *The Anaesthetic Revelation and the Gist of Philosophy* (Amsterdam, NY, 1874), p. 25. References hereafter to *Anaesthetic Revelation* in text.
24 Benjamin Paul Blood, *Pluriverse: An Essay in the Philosophy of Pluralism* (London, 1921), p. 231.
25 James, 'Recent Literature', p. 629.
26 Blood, *Pluriverse*, pp. 241–2.
27 William James, *The Varieties of Religious Experience*, in *The Works of William James*, vol. XV (Cambridge, MA, and London, 1985), p. 308 n.8.
28 James, *The Varieties of Religious Experience*, pp. 307–8.
29 William James, 'A Pluralistic Mystic', in *Memories and Studies* (London, 1910), pp. 374–5.
30 William James, *Pragmatism: A New Name for Some Old Ways of Thinking* (New York, London and Toronto, 1907), p. 50.

four: Gasworks

1 Ralph Waldo Emerson, *Letters and Social Aims* (London, 1976), p. 3.
2 Emerson, *Letters and Social Aims*, p. 4.
3 Malcolm Falkus, 'The British Gas Industry Before 1850', *Economic History Review*, XX (1967), p. 498.
4 John Clayton, 'An Experiment Concerning the Spirit of Coals, Being Part of a Letter to the Hon. Rob. Boyle, Esq.; from the Late Rev. John Clayton, DD Communicated by the Right Rev. Father in God Robert Lord Bishop of Corke to the Right Hon. John Earl of Egmont, FRS John Clayton',

Philosophical Transactions, XLI (1739), p. 60.
5 James Lowther, 'An Account of the Damp Air in a Coal-Pit of Sir James Lowther, Bart. Sunk within 20 Yards of the Sea; Communicated by Him to the Royal Society', *Philosophical Transactions*, XXXVIII (1733), p. 110.
6 Lowther, 'An Account of the Damp Air in a Coal-Pit of Sir James Lowther', p. 112.
7 Charles Hunt, *A History of the Introduction of Gas Lighting* (London, 1907), p. 18.
8 Frederick Albert Winsor, *To Be Sanctioned By An Act of Parliament: A National Light and Heat Company, For Providing Our Streets and Houses With Light and Heat . . .* (London, 1807), p. 12.
9 Frederick Albert Winsor, *A National Light and Heat Company, For Providing Our Streets and Houses With Hydrocarbonic Gas-Lights, On Similar Principles, As They Are Now Supplied With Water . . .* (London, 1805), p. 10.
10 Frederick Albert Winsor, *Plain Questions and Answers. Refuting Every Possible Objection Against the beneficial Introduction of Coke and Gas Lights* (London, 1807), p. 5.
11 Frederick Albert Winsor, *Analogy Between Animal and Vegetable Life. Demonstrating the Beneficial Application of the Patent Light Stoves, To All Green & Hot Houses* (London, 1807), p. 14.
12 Quoted in Dean Chandler, *Outline of History of Lighting by Gas* (London, 1936), p. 24.
13 *An Heroic Epistle to Mr Winsor, The Patentee of the Hydro-Carbonic Gas Lights and Founder of the National Light and Heat Company* (London, 1808), pp. 7, 8.
14 *Lord Granville Leveson Gower (First Earl Granville): Private Correspondence, 1781 to 1821*, ed. Castalia, Countess Granville, 2 vols (London, 1916), vol. II, p. 281.
15 *The Gas-Light Bill* (London, 1809), pp. 10–11.
16 *Heroic Epistle to Mr Winsor*, p. 18.
17 Bruno Latour, *Reassembling the Social: An Introduction to Actor-Net work Theory* (Oxford, 2005), p. 79.
18 Gaston Bachelard, *The Flame of a Candle*, trans. Joni Caldwell (Dallas, 1988), p. 24.
19 Wolfgang Schivelbusch, *Disenchanted Night: The Industrialisation of Light in the Nineteenth Century*, trans. Angela Davies (Oxford, New York and Hamburg, 1988), p. 44.
20 Bachelard, *Flame of a Candle*, p. 64.
21 Bachelard, *Flame of a Candle*, p. 3.
22 'Pamphlets on the Gas-Lights', *Edinburgh Review*, XIII (1809), p. 487.
23 Quoted in Wolfgang Schivelbusch, *Disenchanted Night*, p. 96.
24 Lynda Nead, *Victorian Babylon: People, Streets and Images in Nineteenth-Century London* (New Haven and London, 2000), p. 98.
25 Olive Logan, *The Mimic World, and Public Exhibitions: Their History, Their Morals, and Effects* (Philadelphia and Cincinnati, 1871), p. 132.

26 Quoted in Logan, *The Mimic World*, p. 415.
27 Bachelard, *Flame of a Candle*, p. 69.
28 Bachelard, *Flame of a Candle*, p. 64.
29 Charles Dickens, *The Adventures of Oliver Twist*, ed. Steven Connor (London, 1994), p. 109.
30 Wilkie Collins, *Basil* (New York, 1980), p. 166.
31 George Ellington, *The Women of New York; or, The Under-world of the Great City. Illustrating the Life of Women of Fashion, Women of Pleasure, Actresses and Ballet Girls, Saloon girls, Pickpockets and Shoplifters, Artists' Female Models, Women-of-the-town, etc. . . .* (New York, 1869), p. 344.
32 Mary Elizabeth Braddon, *Lady Audley's Secret*, ed. Jenny Bourne Taylor (London, 1998), p. 395.
33 Edward Bulwer-Lytton, *The Parisians* (London, 1875) vol. I, p. 182.
34 Bulwer-Lytton, *The Parisians*, p. 182.
35 Emerson Bennett, *Ellen Norbury; or, The Adventures of an Orphan* (Philadelphia, 1855), p. 35.
36 Charles Dickens, *The Posthumous Papers of The Pickwick Club*, ed. Robert L. Patten (London, 1972), p. 573.
37 Julian Wolfreys, *The Old Story, with a Difference: Pickwick's Vision* (Columbus, OH, 2006), p. 94.
38 Wolfreys, *The Old Story*, p. 112 n. 7.
39 Andrew Pritchard, 'On the Construction and Management of Solar and Oxy-hydrogen Gas Microscopes, &c', in C. R. Goring and Andrew Pritchard, *Micrographia: Containing Practical Essays on Reflecting, Solar, Oxy-hydrogen Gas Microscopes, Micrometers, Eye-pieces, &c. &c.* (London, 1837), p. 171.
40 George Eliot, *Middlemarch*, ed. W. J. Harvey (Harmondsworth, 1976), p. 177.
41 Zerah Colburn, *The Gas-Works of London* (London, 1865), pp. 24–5.
42 Flora Tristan, *London Journal: A Survey of London Life in the 1830s*, trans. Dennis Palmer and Giselle Pincetl (London, 1980), pp. 67–8.
43 Tristan, *London Journal*, p. 68.
44 Malcolm Falkus, 'The Early Development of the Gas Industry, 1790–1815', *Economic History Review*, XXXV (1982), p. 219.
45 Quoted in Hugh Barty-King New Flame., *How Gas Changed the Commercial, Domestic and Industrial Life of Britain between 1813 and 1984* (Tavistock, 1984), p. 257.
46 Thomas C. Hine, *Warming and Ventilation* (London, 1874), p. 6.
47 Dean Chandler, *Outline of History of Lighting by Gas*, p. 160.
48 Nead, *Victorian Babylon*, p. 102.
49 Winsor, *Plain Questions and Answers*, pp. 5–6.
50 *Heroic Epistle to Mr Winsor*, p. 3.
51 Malcolm Falkus, *Always under Pressure: History of North Thames Gas Since 1949* (Basingstoke, 1988), p. 11.
52 Ellington, *Women of New York*, p. 112.

53 Dickens, *Pickwick Papers*, p. 86.
54 'Gas Nuisances and Their Removal', *Sanitary Review and Journal of Public Health*, III (1857), pp. 191–2.
55 Evariste Bertulus, *Mémoire d'hygiène publique sur cette question: Rechercher l'influence que peut exercer l'éclairage au gaz sur la santé des masses dans l'intérieur des villes?* (Marseilles, 1853).
56 Thomas Newbigging, 'The Gas Industry of the United Kingdom', *Scottish Review*, IX (1887), p. 307.
57 'Suicide and Suggestion.' *The Lancet*, CLXXII/4440 (3 October 1908), p. 1041.
58 Olive Anderson, *Suicide in Victorian and Edwardian Britain* (Oxford, 1987), p. 352.
59 George Augustus Sala, 'The Secrets of the Gas', *Gaslight and Daylight: Some London Scenes They Shine Upon* (London, 1859), p. 163.
60 Sala, 'The Secrets of the Gas', p. 159.
61 Charles Dickens, *Bleak House*, ed. Norman Page (Harmondsworth, 1972), pp. 49–50.
62 Dickens, *Bleak House*, p. 851.
63 Dickens, *Bleak House*, p. 422.
64 'Pamphlets on the Gas-Lights', p. 478.
65 Dickens, *Bleak House*, p. 683.
66 Sala, 'The Secrets of the Gas', p. 159.
67 Dickens, *Bleak House*, p. 683.
68 Latour, *Reassembling the Social*, pp. 37–42.
69 Latour, *Reassembling the Social*, 46.
70 Latour, *Reassembling the Social*, pp. 2, 9.
71 Latour, *Reassembling the Social*, p. 89.

five: Transported Shiver of Bodies: Weighing Ether

1 Thomas Young, 'Experiments and Calculations Relative to Physical Optics', *Philosophical Transactions of the Royal Society*, XCIV (1804), pp. 12–13.
2 Gillian Beer, 'Wave Theory and the Rise of Literary Modernism', in *Open Fields: Science in Cultural Encounter* (Oxford, 1996), p. 298.
3 William Thomson, Baron Kelvin, *Popular Lectures and Addresses*, 3 vols (London, 1891), vol. I, p. 310.
4 Albert Einstein, 'The History of Field Theory: Olds and News of Field Theory', *New York Times* (3 February 1929). Online at www.rain.org/~karpeles/einsteindis.html <http://www.rain.org/~karpeles/einsteindis.html> (accessed 26 August 2010).
5 Alfred Senior Merry, *Interstellar Aether* (London, 1891), pp. 6–7.
6 Hermann Helmholtz, 'Über Integrale der hydrodynamische Gleichungen, welche den Wirbelbewegungen entsprechen', *Journal für die Reine und Angewandte Mathematik*, LV (1858), pp. 25–55.
7 William Thomson, 'On Vortex Atoms', *Philosophical Magazine*, XXXIV (1867), pp. 15–24.

8 James Clerk Maxwell, *The Scientific Papers of James Clerk Maxwell*, ed. W. D. Niven, 2 vols (Cambridge, 1890), vol. II, p. 247.
9 Helge Kragh, 'The Vortex Atom: A Victorian Theory of Everything', *Centaurus*, XLIV (2002), pp. 32–114.
10 John Wills Cloud, *The Ether and Growth: A Theoretical Study* (London, 1928), p. 9.
11 Cloud, *The Ether and Growth*, p. 20.
12 John Tyndall, *Fragments of Science for Unscientific People* (London, 1871), p. 163. References hereafter to *Fragments of Science* in text.
13 William Whewell, *Astronomy and General Physics Considered With Reference to Natural Theology* (London, 1833), pp. 138–9.
14 Whewell, *Astronomy and General Physics Considered With Reference to Natural Theology*, pp. 139, 140.
15 Whewell, *Astronomy and General Physics Considered With Reference to Natural Theology*, p. 140.
16 Peter J. Bowler, *Reconciling Science and Religion: The Debate in Early Twentieth-Century Britain* (Chicago and London, 2001), p. 89.
17 Oliver Lodge, *Modern Views on Matter* (Oxford, 1907), p. 23.
18 Joseph Conrad, *Collected Letters*, vol. II, ed. Frederick R. Karl and Laurence Davies 9 vols (Cambridge, 1983–2007), vol. II, pp. 94–5.
19 Ernst Haeckel, *The Riddle of the Universe at the Close of the Nineteenth Century*, trans. Joseph McCabe (London, 1900); John Sloan, *John Davidson, First of the Moderns: A Literary Biography* (Oxford, 1995), p. 215; John Davidson, *The Triumph of Mammon* (London, 1907), p. 117.
20 G. N. Cantor, 'The Theological Significance of Ethers', in *Conceptions of Ether: Studies in the History of Ether Theories, 1740–1900*, ed. G. N. Cantor and M.J.S. Hodge (Cambridge, 1981), pp. 135–55.
21 Conrad, *Collected Letters*, vol. II, p. 95.
22 Arthur Balfour, *Reflections Suggested By the New Theory of Matter* (London, 1904), pp. 9, 21.
23 Balfour, *Reflections Suggested By the New Theory of Matter*, p. 18.
24 Balfour, *Reflections Suggested By the New Theory of Matter*, p. 10.
25 Balfour, *Reflections Suggested By the New Theory of Matter*, p. 10.
26 Walter Pater, *The Renaissance: Studies in Art and Poetry*, ed. Adam Phillips (Oxford, 1998), pp. 150, 152.
27 Ralph Waldo Emerson, *Essays: Second Series* (London, 1844), p. 45.
28 Pater, *The Renaissance*, p. 152.
29 Roger K. French, 'Ether and Physiology', in *Conceptions of Ether*, ed. Cantor and Hodge, pp. 111–34.
30 David Hartley, *Observations on Man, His Frame, His Duty, and His Expectations*, 2 vols (London, 1749), vol. I, pp. 13–15.
31 James Stanley Grimes, *Etherology, and the Phreno-philosophy of Mesmerism and Magic Eloquence: Including a New Philosophy of Sleep and of Consciousness . . .*, 2nd edn, ed. W. G. Le Duc (Boston, MA, 1850).
32 Benjamin Ward Richardson, 'The Theory of a Nervous Ether', in *Half-Hour Recreations in Popular Science*, 1st series, ed. Diana Estes (Boston, MA,

1874), p. 368.
33 Richardson, 'The Theory of a Nervous Ether', pp. 368–9.
34 Quoted in John Barrow, *The Book of Nothing* (London, 2001), p. 130.
35 Jean Lorrain, *Nightmares of an Ether Drinker*, trans. Brian Stableford (Oxford, 2002).
36 Richardson, 'Theory of a Nervous Ether', p. 364.
37 Richardson, 'The Theory of a Nervous Ether', p. 372.
38 Claus Priesner, 'Spiritus Aethereus: Formation of Ether and Theories of Etherification From Valerius Cordus to Alexander Williamson', *Ambix*, XXXIII (1986), pp. 129–52.
39 Matthew Turner, *An Account of the Extraordinary Medicinal Fluid, Called Aether* (London, 1761), p. 4.
40 Turner, *An Account of the Extraordinary Medicinal Fluid*, pp. 4–5.
41 Quoted in Mike Jay, *Emperors of Dreams: Drugs in the Nineteenth Century* (Sawtry, 2000), p. 142.
42 Jay, *Emperors of Dreams*, p. 142.
43 T. E. Hulme, *The Collected Writings of T. E. Hulme*, ed. Karen Csengeri (Oxford, 1994), pp. 62–3.
44 Christopher Milbourne, *The Illustrated History of Magic* (London, 1975), p. 145.
45 I. Bernard Cohen, 'The First English Version of Newton's *Hypotheses non fingo*', *Isis*, LII (1962), pp. 379–88.
46 Samuel Taylor Coleridge, *Aids to Reflection* (London, 1825), p. 394.
47 Basil Mahon, *The Man Who Changed Everything: The Life of James Clerk Maxwell* (Hoboken, NJ, 2003), p. 3.
48 Quoted in Silas W. Holman, *Matter, Energy, Force and Work: A Plain Presentation of Fundamental Physical Concepts and of the Vortex Atom and Other Theories* (New York, 1898), p. 226.
49 Herbert Spencer, *An Autobiography*, 2 vols (London, 1904), vol. II, p. 19.
50 David Duncan, *The Life and Letters of Herbert Spencer* (London, 1911), p. 550.
51 Herbert Spencer, *The Principles of Psychology* (New York, 1898), p. 619.
52 Spencer, *The Principles of Psychology*, pp. 619–20.
53 Rudyard Kipling, *Traffics and Discoveries*, ed. Hermione Lee (Harmondsworth, 1987), pp. 181–99.

six: Haze

1 Salman Rushdie, *The Satanic Verses* (London, 1988), pp. 354–5.
2 Virginia Woolf, *Orlando: A Biography* (London, 1972), p. 157.
3 Woolf, *Orlando*, pp. 158–9.
4 Michel Serres, *The Natural Contract*, trans. Elizabeth MacArthur and William Paulson (Ann Arbor, MI, 1995), p. 4.
5 Steven Connor, *The Book of Skin* (London, 2003), p. 159.
6 Matthew Arnold, *Poems*, ed. Kenneth Allott, 2nd edn., ed. Miriam Allott (London, 1979), pp. 259–60.

7 Alexander Pope, *The Correspondence of Alexander Pope*, ed. George Sherburn, 5 vols (Oxford, 1956), vol. II, p. 522.
8 Charles Dickens, *Bleak House*, ed. Norman Page (Harmondsworth, 1972), p. 49.
9 Ezra Pound, 'A Few Don'ts By An Imagiste;, *Poetry* I (1913), p. 201.
10 Wyndham Lewis et al., 'Manifesto', *Blast: Review of the Great English Vortex*, I (1914), p. 11.
11 Lewis, 'Manifesto', p. 12.
12 Eugene Umberger, 'In Praise of Lady Nicotine: A Bygone Era of Prose, Poetry . . . and Presentation', in *Smoke: A Global History of Smoking*, ed. Sander L. Gilman and Zhou Xun (London, 2004), pp. 236–47.
13 Arthur Conan Doyle, *The Penguin Complete Sherlock Holmes* (London, 1981), p. 240.
14 Lydia Marinelli, 'Smoking, Laughing and the Compulsion to Film: On the Beginnings of Psychoanalytic Documentaries', *American Imago*, LXI (2004), p. 35.
15 T. S. Eliot, *Inventions of the March Hare: Poems, 1909–1917*, ed. Christopher Ricks (London, 1996), p. 51.
16 Eliot, *Inventions of the March Hare*, p. 70.
17 Gaston Bachelard, *Air and Dreams: An Essay on the Imagination of Movement*, trans. Edith R. Farrell and C. Frederick Farrell (Dallas, TX, 1988), p. 127.
18 Friedrich Nietzsche, *Untimely Meditations*, trans. R. J. Hollingdale (Cambridge, 1983), pp. 64–5, 95, 97.
19 Nietzsche, *Untimely Meditations*, pp. 94–5.
20 Virginia Woolf, 'Modern Fiction', in *The Common Reader: First Series* (London, 1984), p. 150.
21 Paul Verlaine, *Oeuvres poétiques complètes* (Paris, 1962), p. 362.
22 Michel Serres, *Genesis*, trans. Geneviève James and James Nielson (Ann Arbor, MI, 1995), p. 19.
23 Honoré de Balzac, 'Le Chef-d'oeuvre inconnu', in *La Comédie humaine*, vol. x (Paris, 1979), p. 436 [my translation].
24 Balzac, 'Le Chef-d'oeuvre inconnu', p. 435.
25 Serres, *Genesis*, pp. 18–19.
26 Joseph Conrad, *The Nigger of the 'Narcissus': A Tale of the Sea*, ed. Jacques Berthoud (Oxford and New York, 1984), p. xlii.
27 Joseph Conrad, *Heart of Darkness*, 2nd edn, ed. Robet Kimbrough (New York, 1971), p. 5.
28 Conrad, *The Nigger of the 'Narcissus'*, p. 4.
29 Conrad, *The Nigger of the 'Narcissus'*, p. 40.
30 Martine Hennard Dutheil de la Rochère, 'Body Politics: Conrad's Anatomy of Empire in Heart of Darkness', *Conradiana*, XXXVI (2004), pp. 185–205; Conrad, *Heart of Darkness*, p. 66.
31 Conrad, *Heart of Darkness*, pp. 40–41.
32 John Tyndall, *Essays on the Floating Matter of the Air, in Relation to Putrefaction and Infection* (London, 1881), p. xiii.

33 Elizabeth Diller, *Blur: The Making of Nothing* (New York, 2002).
34 Richard Hamblyn, 'A Celestial Journey', *Tate Etc*, v (2005), p. 90.
35 Usman Haque, 'Sky Ear: Concepts and Final Design' (2004), p. 1. Online at www.haque.co.uk/skyear/skyearconceptsanddesign.pdf (accessed 12 April 2010).
36 Usman Haque, 'Floatables', (2009). Online at www.haque.co.uk/floatables.php (accessed 12 April 2010).
37 Woolf, 'Modern Fiction', p. 150.
38 Haque, 'Floatables'.
39 Conrad, *Heart of Darkness*, p. 40.

seven: Atmospherics

1 Thomas De Quincey, *De Quincey as Critic*, ed. John Jordan (London and Boston, MA, 1973), p. 404.
2 W. E. Ayrton, 'Sixty Years of Submarine Telegraphy', *The Electrician*, XLII (19 February 1897), p. 548.
3 Samuel Beckett, *Complete Dramatic Works* (London, 1986), p. 96.
4 'A Unique Record', *Marconigraph*, II (1912), p. 275.
5 William Crookes, 'Some Possibilities of Electricity', *Fortnightly Review*, n.s., LI (1892), p. 175.
6 Walter Benjamin, 'The Work of Art in the Age of Mechanical Reproduction', *Illuminations*, ed. Hannah Arendt, trans. Harry Zohn (London, 1969), p. 239.
7 Oliver Lodge, *Talks About Wireless: With Some Pioneering History and Some Hints and Calculations for Wireless Amateurs* (London, 1925), p. 45.
8 'The Electric Wave', *New York Times* (12 January, 1873), p. 3. Online at www.earlyradiohistory.us/1873wave.htm (accessed 12 April 2010).
9 H. Morris Airey and W. H. Eccles, 'Note on The Electrical Waves Occurring in Nature', *Proceedings of the Royal Society*, LXXXV (1911), p. 146.
10 'Cantab', 'Strays and their Origin', *Wireless World*, VIII (1920), pp. 346–7.
11 R. A. Watson Watt, 'Observations on Atmospherics', *Wireless World*, XII (1923), p. 601.
12 'Discovery in Wireless: "Atmospherics" Eliminated', *The Times*, (8 March 1919), p. 11.
13 'The Weagant "x-Stopper"', *Wireless World*, VII (1919), p. 127.
14 Albert Turpain, *La Prévision des orages* (Paris, 1902).
15 Airey and Eccles, 'Note on The Electrical Waves', p. 150.
16 J. Reginald Allinson, 'Atmospherics', *Weather and Wireless Magazine*, II (1924), p. 14.
17 A.G. McAdie, 'Volcanic Eruptions and their Effect upon Wireless Telegraphy', *Marconigraph*, II (1913), p. 490.
18 Carl Størmer, 'Short Wave Echoes and the Aurora Borealis', *Nature*, CXXII (1928), p. 681.
19 S. Chapman, 'The Audibility and Lowermost Altitude of the Aurora Polaris', *Nature* CXXVII, (1931), p. 342.

20 Oliver Lodge, *Signalling across Space Without Wires*. 4th edn (London, 1908), p. 33.
21 'No Message From Mars: Marconigrams Into Space', *The Times*, (24 April 1920), p. 14.
22 Quoted in Ron, 'Mars Calling Earth', *Radio Bygones*, LXXIV (2001), p. 5.
23 Quoted in Ron, 'Mars Calling Earth', p. 4.
24 *Wireless World*, VIII (1920), p. 101.
25 Karl G. Jansky, 'Electrical Disturbances Apparently of Extraterrestrial Origin', *Proceedings of the Institute of Radio Engineers*, XXI (1933), pp. 1387–98.
26 Bernard Lovell, *Voice of the Universe: Building the Jodrell Bank Telescope* (London, 1987), p. 21.
27 'Commendatore G. Marconi', *Wireless World*, I (1913), p. 3.
28 J. J. Fahie, *A History of Wireless Telegraphy, 1838–1899* (Edinburgh and London, 1899), p. 80.
29 'The Weagant "x-Stopper"', p. 129.
30 'Atmospherics. Wireless Forecasts of Thunderstorms', *Times*, 43984 (10 June 1925), p. 8.
31 Heinrich Barkhausen, 'Pfeiftöne aus der Erde', *Physikalische Zeitschrift*, XX (1919), pp. 402–3 [my translation].
32 Heinrich Barkhausen, 'Whistling Tones From the Earth', *Proceedings of the Institute of Radio Engineers*, XVIII (1930), pp. 1155–9.
33 'Morse Signals: A Growing Source Of Interference', *The Times*, (24 June 1926), p. 28.
34 E. T. Burton and E. M. Boardman, 'Audio-Frequency Atmospherics', *Proceedings of the Institute of Radio Engineers*, XXI (1933), pp. 1481, 1487.
35 'Amateur Notes', *Wireless World*, I (1913), p. 587.
36 E. W. Marchant, 'Methods of Reducing Interference in Wireless Receiving Sets: Discussion', *Wireless World*, XII (1923), p. 465.
37 B.C.L., 'Wireless Freaks', *The Times*, (19 January 1926), p. 17.
38 'Strange Causes of Interference: Unsuspected Crystals', *The Times*, (26 October 1926), p. 24.
39 Paul Deresco Augsburg, *On The Air* (New York, 1927), p. 4.
40 Alvin F. Harlow, *Old Wires and New Waves: The History of the Telegraph, Telephone, and Wireless* (New York and London, 1936), p. 449.
41 Sungook Hong, *Wireless: From Marconi's Black-Box to the Audion* (Cambridge, MA, and London, 2001), pp. 107–12.
42 J. A. Fleming, 'Wireless Telegraphy at the Royal Institution', *The Times*, (11 June 1903), p. 4.
43 Nevil Maskelyne, 'Wireless Telegraphy', *Times*, 37106 (13 June 1903), p. 9.
44 Quoted in Helen M. Fessenden, *Fessenden: Builder of Tomorrows* (New York, 1940), p. 121.
45 Patrick Vaux, *Sea-Salt and Cordite* (London, New York and Toronto, 1914), p. 47.
46 W. E. Collinson, *Contemporary English: A Personal Speech Record* (Leipzig and Berlin, 1927), p. 113.

47 J. Reginald Allinson, 'Tracking Thunderstorms with a Frame Aerial', *Weather and Wireless Magazine*, II (1924), p. 50.
48 Mark R. Rosenzweig and Geraldine Stone, 'Wartime Research in Psycho-Acoustics', *Review of Educational Research*, XVIII (1948), pp. 642–54.
49 Hugh G. J. Aitken, *Syntony and Spark: The Origins of Radio* (New York and London, 1976), pp. 33, 37.
50 F. T. Marinetti and Pino Masnata, 'La Radia', in *Wireless Imagination: Sound, Radio and the Avant-Garde*, ed. Douglas Kahn and Gregory Whitehead (Cambridge, MA, and London, 1992), p. 267.
51 Rupert T. Gould, 'The History of the Chronometer', *Geographical Review*, LVII (1921), p. 269.
52 W. H. Eccles, *Atmospheric Telegraphy and Telephony: A Handbook of Formulae, Data and Information*, 2nd edn (London 1918), p. 177.
53 Eccles, *Atmospheric Telegraphy and Telephony*, p. 164.
54 Quoted in Albert Glinsky, *Theremin: Ether Music and Espionage* (Urbana and Chicago, 2000), p. 71. References hereafter to *Theremin* in text.
55 Siegfried Sassoon, *Collected Poems, 1908–1956* (London, 1984), p. 238.
56 Sassoon, *Collected Poems*, pp. 238–9.
57 Friedrich Jürgenson, *Radio- och mikrofonkontakt med de döda* (Uppsala, 1968); Konstantin Raudive, *Breakthrough: An Amazing Experiment in Electronic Communication with the Dead*, trans. Nadia Fowler, ed. Joyce Morton (Gerrards Cross, 1971).
58 Edfrid A. Bingham and John Parslow, 'Fighting Fog with Hertzian Waves', *Technical World Magazine* (July 1909), pp. 520–23. Online at www.earlyradiohistory.us/1909fog.htm (accessed 12 April 2010).
59 J. Patrick Wilson, 'The Technological Heritage of Oliver Lodge', in *Oliver Lodge and the Invention of Radio*, ed. Peter Rowlands and J. Patrick Wilson (Liverpool, 1994), pp. 189–90.
60 Quoted in F. Graham Smith, *Radio Astronomy*, 4th edn (Harmondsworth, 1974), p. 18.
61 Smith, *Radio Astronomy*, p. 29.
62 L. C. Hall, 'Telegraph Talk and Talkers: Human Character and Emotions an Old Telegrapher Reads on the Wire', *McClure's Magazine*, XVIII (1902), p. 227.
63 Hall, 'Telegraph Talk and Talkers', p. 229.
64 Hall, 'Telegraph Talk and Talkers', p. 231.
65 Arthur P. Harrison Jr, 'Single-control Tuning: An Analysis of An Innovation', *Technology and Culture*, XX (1979), pp. 296–321.
66 Jonathan Hill, *The Cat's Whisker: 50 Years of Radio Design* (London, 1978), p. 77.
67 Harrison Jr, 'Single-control Tuning', pp. 73–9.
68 Lord Dunsany, *Plays for Earth and Air* (London and Toronto, 1937), p. 155.
69 Dunsany, *Plays for Earth and Air*, p. 163.
70 Dunsany, *Plays for Earth and Air*, p. 110.
71 Arthur L. Salmon, 'Is Broadcasting a Disaster?' *Musical Times*, LXVI

(1 September 1925), p. 796.
72 'Auribus', 'Wireless Notes', *Musical Times*, LXX (1 December 1929), p. 1095.
73 'Auribus', 'Wireless'.
74 Sassoon, *Collected Poems*, p. 158.
75 Quoted in Fahie, *History of Wireless Telegraphy*, pp. 42–3.
76 '"Wireless Music": A Novel Invention', *The Times*, (8 December 1927), p. 15.
77 Andrea Polli, 'Modelling Storms in Sound: The Atmospherics/Weather Works Project', *Organised Sound*, IX (2004), pp. 175–80; '*Atmospherics/Weather Works*: A Spatialized Meteorological Data Sonification Project', *Leonardo*, XXXVIII (2005), pp. 31–6.
78 David Toop, *Haunted Weather: Music, Silence and Memory* (London, 2004), p. 100.
79 Aristides Quintilianus, *On Music: In Three Books*, ed. and trans. Thomas J. Mathiesen (New Haven and London, 1983), p. 152.

eight: A Grave in the Air

1 Sophocles, *Antigone*, trans. Paul Woodruff (Indianapolis, 2001), ll. 418–22, p. 18.
2 Sophocles, *Antigone*, ll. 1005–11, pp. 43–4.
3 Dina Rabinovitch, 'His Bright Materials', *Guardian*, *G2* (10 December 2003), p. 14.
4 Tim Radford, 'Ultimate Trip as LSD Guru Leary is Truly Spaced Out', *Guardian* (21 April 1997), p. 2; Jane Cassidy, 'He Went Out With A Bang', *New Review* (23 December 2007), p. 32.
5 Katherine Q. Seelye, 'Ashes-to-Fireworks Send-Off for an "Outlaw" Writer', *New York Times* (22 August 2005), section A, p. 8.
6 John Keats, *Poetical Works*, ed. H. W. Garrod, 2nd edn (Oxford, 1958), p. 259.
7 Edward Trelawny, *Recollections of the Last Days of Shelley and Byron* (London, 2000), pp. 91–2.
8 V.A.C. Gatrell, *The Hanging Tree: Execution and the English People, 1770–1868* (Oxford, 1994), p. 337.
9 Henry Thompson, 'The Treatment of the Body after Death', *Contemporary Review*, XXIII (1874), p. 327. References hereafter to 'Treatment of the Body' in text.
10 Quoted in Stephen Prothero, *Purified By Fire: A History of Cremation in America* (Berkeley, CA, and London, 2001), pp. 93–4.
11 Quoted in Robert Nicol, *This Grave and Burning Question: A Centenary History of Cremation in Australia* (Adelaide, 2003), p. 53.
12 Laurence Binyon, *The North Star and Other Poems* (London, 1941), p. 59.
13 Henry Thompson, 'Cremation: A Reply to Critics and an Exposition of the Process', *Contemporary Review*, XXIII (1874), p. 561. References hereafter to 'Cremation' in text.
14 George Wotherspoon, *Cremation, Ancient and Modern: The History and Utility of Fire-Funeral* (London, 1886), p. 12.

15 Henry Thompson, *Modern Cremation: Its History and Practice to the Present Date*, 4th edn (London, 1901), p. 147.
16 Prothero, *Purified By Fire*, p. 149.
17 Wotherspoon, *Cremation, Ancient and Modern*, p. 15.
18 Philip H. Holland, 'Burial or Cremation? A Reply to Sir Henry Thompson', *Contemporary Review*, XXIII (1874), p. 483.
19 Quoted in Prothero, *Purified By Fire*, p. 53.
20 Robert Browning, *Poetical Works*, ed. Ian Jack, Rowena Fowler and Margaret Smith, vol. IV (Oxford, 1991), p. 73.
21 Elizabeth Bloch-Smith, *Judahite Burial Practices and Beliefs about the Dead* (Sheffield, 1992), pp. 52, 54n.
22 Peter Jupp, *From Dust to Ashes: The Replacement of Burial by Cremation in England, 1840–1967* (London, 1978), p. 16.
23 John Twigg, 'New Light on the Ashes', *International Journal of the History of Sport*, 4 (1987), pp. 231–6.
24 Mrs Howard Vincent, *Forty Thousand Miles Over Land and Water: The Journal of a Tour Through the British Empire and America*, 2 vols (London, 1886), vol. I, p. 233.
25 Vincent, *Forty Thousand Miles Over Land and Water*, vol. I, p. 234.
26 Salman Rushdie, *The Satanic Verses* (London, 1988), p. 5.
27 Curt Wachtel, *Chemical Warfare* (London, 1941), p. 21.
28 Siegfried Sassoon, *Memoirs of an Infantry Officer* (London, 1930), p. 147.
29 F. N. Pickett, *Don't Be Afraid of Poison Gas: Hints for Civilians in the Event of a Poison Gas Attack* (London, 1934), p. 18.
30 Robert Harris and Jeremy Paxman, *A Higher Form of Killing: The Secret History of Gas and Germ Warfare*, 2nd edn (London, 2002), p. 8.
31 Roger Highfield, 'Blood Tests Confirm Gas Was Common Anaesthetic', *Daily Telegraph* (30 October 2002), p. 15.
32 Quoted in Anon., *Poison Gas* (London, 1935), p. 8.
33 Harris and Paxman, *Higher Form of Killing*, p. 3.
34 Wilfred Owen, *War Poems and Others*, ed. Dominic Hibbert (London, 1976), p. 79.
35 Harris and Paxman, *Higher Form of Killing*, p. 87.
36 L. F. Haber, *The Poisonous Cloud: Chemical Warfare in the First World War* (Oxford, 1986), p. 275.
37 Harris and Paxman, *Higher Form of Killing*, pp. 100–01.
38 Quoted in Harris and Paxman, *Higher Form of Killing*, p. 111.
39 Quoted in Harris and Paxman, *Higher Form of Killing*, p. 108.
40 William Bache, *An Inaugural Experimental Dissertation, Being an Endeavour to Ascertain the Morbid Effects of Carbonic Acid Gas; or, Fixed Air, on Healthy Animals, and the Manner in Which They Are Produced* (Philadelphia, 1794), p. 45.
41 Bache, *An Inaugural Experimental Dissertation*, p. 46.
42 Quoted in James Kendall, *Breathe Freely! The Truth about Poison Gas* (London, 1938), p. 74.
43 Pickett, *Don't Be Afraid of Poison Gas*, p. 12.

44 Anon., *Poison Gas*, p. 22.
45 Anon., *Poison Gas*, p. 28.
46 Quoted in Harris and Paxman, *Higher Form of Killing*, p. 189.
47 Bruno Latour, We Have Never Been Modern, trans. Catherine Porter (Hemel Hempstead, 1993), pp. 13–18, 51–5.
48 Camille Paglia, *Sexual Personae: Art and Decadence From Nefertiti to Emily Dickinson* (New Haven, 1990), p. 21.
49 Peter Bamm, *Eines Menschen Zeit* (Zurich, 1972), p. 320.
50 Bruno Latour, *We Have Never Been Modern*, trans. Catherine Porter (New York and London, 1993), p. 77.
51 Latour, *We Have Never Been Modern*, p. 78.
52 Steven Connor, 'Michel Serres's Milieux' (2002). Online at www.stevenconnor.com/milieux
53 Owen, *War Poems and Others*, p. 79.
54 Quoted in Bryan Vila and Cynthia Morris, *Capital Punishment in the United States: A Documentary History* (London and Westport, CT, 1997), p. 78.
55 Harris and Paxman, *Higher Form of Killing*, p. 241.
56 Anon., *Poison Gas*, p. 9.
57 Michel Serres, *La Guerre mondiale* (Paris, 2008), pp. 47–92.
58 Peter Sloterdijk, *Schäume: Sphären, vol. 3: Plurale Sphärologie* (Frankfurt, 2004),
59 M. H. Pickering, *Air Poems and Others* (Ilfracombe, 1947), p. 5.
60 Binyon, *North Star*, p. 19.
61 W. G. Sebald, *On The Natural History of Destruction*, trans. Anthea Brell (London, 2003), p. 27.
62 H. G. Wells, *The War in the Air: And Particularly How Mr Bert Smallways Fared While It Lasted* (London, 2002), p. 340.
63 E. H. Horne, *The Significance of Air War: An Essay in Interpretation* (London and Edinburgh, 1937), p. 10.
64 David Gascoyne, *Selected Poems* (London, 1994), p. 135.
65 HD, *Trilogy: The Walls Do Not Fall; Tribute to the Angels; The Flowering of the Rod* (Cheadle Hulme, 1973), pp. 58–9.
66 Gerard Manley Hopkins, *The Poems*, 4th edn, ed. W. H. Gardner and N. H. Mackenzie (London, 1970), p. 19.
67 T. S. Eliot, *The Complete Poems and Plays of T. S. Eliot* (London, 1969), p. 192.
68 Eliot, *The Complete Poems and Plays of T. S. Eliot*, p. 193.
69 Eliot, *The Complete Poems and Plays of T. S. Eliot*, p. 198.
70 Eliot, *The Complete Poems and Plays of T. S. Eliot*, p. 196.
71 Edith Sitwell, *Collected Poems* (London, 1993), p. 272.
72 Dylan Thomas, *The Poems*, ed. Daniel Jones (London and Melbourne, 1985), p. 175.
73 Thomas, *The Poems*, p. 175.
74 Thomas, *The Poems*.
75 Thomas, *The Poems*, p. 172.

76 Thomas, *The Poems*, pp. 172–3.
77 Henry Reed, *Collected Poems*, ed. Jon Stallworthy (Oxford, 1991), p. 15.
78 Paul Celan, *Selected Poems*, trans. Michael Hamburger (London, 1990), pp. 60–63 [my translation].
79 *Air Force Poetry*, eds John Pudney and Henry Treece (London, 1944), p. 58.
80 Celan, *Selected Poems*, pp. 144–5 (translation modified).
81 Roy Fuller, *New and Collected Poems* (London, 1985), p. 42.
82 Martin Amis, *Time's Arrow; or, The Nature of the Offence* (London, 1991), p. 128.
83 Amis, *Time's Arrow*, p. 128.
84 Amis, *Time's Arrow*, p. 145.
85 Walter Benjamin, *Illuminations*, ed. Hannah Arendt, trans. Harry Zohn (London, 1969), pp. 257–8.
86 Luce Irigaray, *The Forgetting of Air in Martin Heidegger*, trans. Mary Beth Mader (London, 1999), pp. 8, 40.
87 Irigaray, *The Forgetting of Air in Martin Heidegger*, p. 9.
88 Lancelot Andrewes, XCVI. *Sermons by the Right Honorable and Reverend Father in God, Lancelot Andrewes, late Lord Bishop of Winchester* (London, 1629), p. 584.
89 Andrewes, XCVI.
90 Nathaniel Heywood, *Christ Displayed as the Choicest Gift, and Best Master . . .* (London, 1679), p. 118.
91 Jean-Baptiste van Helmont, *A Ternary of Paradoxes: The Magnetick Cure of Wounds, Nativity of Tartar in Wine, Image of God in Man*, trans. Walter Charleton (London, 1650), p. 15.
92 Sara Jeanette Duncan, *An American Girl in London* (London, 1891); quoted in Peter Brimblecombe, *The Big Smoke: A History of Air Pollution in London since Medieval Times* (London, 1987), pp. 85–6.
93 Robert Barr, 'The Doom of London', *The Face and the Mask* (London, 1894), p. 79.
94 Barr, 'The Doom of London', p. 88.
95 Barr, 'The Doom of London', p. 90.
96 Arthur Conan Doyle, *The Poison Belt: Being An Account of Another Adventure of Professor George E. Challenger, Lord John Roxton, Professor Summerlee, and Mr. E. D. Malone, the Discoverers of The Lost World* (Leipzig, 1913), pp. 96–7.
97 Conan Doyle, *The Poison Belt*, p. 21.
98 Conan Doyle, *The Poison Belt*, pp. 108-9.
99 Conan Doyle, *The Poison Belt*, pp. 107–8.
100 Conan Doyle, *The Poison Belt*, pp. 231–2.
101 Samuel Beckett, *Complete Dramatic Works* (London, 1986), p. 83.

nine: Air's Exaltation

1 Samuel Sturmy, *The Mariners Magazine; or, Sturmy's Mathematical and Practical Arts* (London, 1669), Book V, p. 80.

2 Quoted in G. I. Brown, *The Big Bang: A History of Explosives* (Stroud, 2005), p. 48.
3 John Bate, *The Mysteries of Nature and Art in Four Severall Parts* (London, 1654), p. 84.
4 Julian Barnes, 'The Art of Suffering', *The Guardian*, 11 May 2002. Online at www.guardian.co.uk/saturday_review/story/0,3605,713363,00.html (accessed 12 April 2010).
5 William Garrard, *Art of Warre* (London, 1591), p. 217.
6 Robert Anderson, *To Hit a Mark, As Well Upon Ascents and Descents, as Upon the Plain of the Horizon Experimentally and Mathematically Demonstrated* (London, 1690), p. 27.
7 Anderson, *To Hit a Mark*.
8 Francis Malthus, *A Treatise of Artificial Fire-works Both for Warres and Recreation . . .* , trans. Thomas Cecil (London, 1629), p. 10.
9 Malthus, *A Treatise of Artificial Fire-works Both for Warres and Recreation*, pp. 14–15.
10 Joseph Moxon, *An Epitome of the Whole Art of War*, 2nd edn (London, 1692), pp. 64–5.
11 Richard Baxter, *Richard Baxter's Catholick Theologie Plain, Pure, Peaceable, for Pacification of the Dogmatical Word-warriours . . .* (London, 1675), p. 88.
12 Richard Lovelace, 'A Fly About a Glasse of Burnt Claret', in *Lucasta Posthume: Poems of Richard Lovelace* (London, 1659), p. 36.
13 Susanna Centlivre, *Mar-plot* (London, 1711), p. 25.
14 Henry More, *The Immortality of the Soul . . .* (London, 1659), p. 464.
15 Margaret Cavendish, *Observations Upon Experimental Philosophy To Which is Added The Description of a New Blazing World* (London, 1666), p. 43.
16 Henry Neville, *The Isle of Pines; or, A Late Discovery of a Fourth Island Near Terra Australis Incognita by Henry Cornelius van Sloetten* (London, 1668), p. 19.
17 Edward Phillips, *The New World of English Words; or, A General Dictionary Containing the Interpretations of Such Hard Words As Are Derived from Other Languages . . .* (London, 1658), n.p.; Elisha Coles, *An English Dictionary* (London, 1677), n.p.
18 W. L. Alden, *Domestic Explosives and Other Sixth Column Fancies* (New York, 1877), p. 7.
19 Ian D. Rae and James H. Whitehead, 'Rackarock: On the Path From Black Powder to ANFO', in *Gunpowder, Explosives and the State: A Technological History*, ed. Brenda J. Buchanan (Aldershot and Burlington, VT, 2006), p. 368.
20 Jacob E. Schmidt, *Narcotics Lingo and Lore* (Springfield, IL, 1959), p. 17.
21 Quoted in Norman Youngblood, *The Development of Mine Warfare: A Most Murderous and Barbarous Conduct* (Westport, CT, and London, 2006), p. 5.
22 Thomas Willis, *An Essay of the Pathology of the Brain and Nervous Stock*

in Which Convulsive Diseases Are Treated of . . ., trans. Samuel Pordage (London, 1681), p. 2.
23 Willis, *An Essay of the Pathology of the Brain . . .*
24 Willis, *An Essay of the Pathology of the Brain . . .*
25 Quoted in Willis, *An Essay of the Pathology of the Brain . . .*, p. 3.
26 Randle Cotgrave, *A Dictionarie of the French and English Tongues . . . Reproduced From the First Edition, London 1611* (Columbia, SC, 1968), n.p.
27 Robert Ward, *Animadversions of War* (London, 1639), p. 113.
28 Ward, *Animadversions of War*, p. 362.
29 William Shakespeare, *Hamlet: The Texts of 1603 and 1623*, ed. Ann Thompson and Neil Taylor (London, 2006), III.4, 208–9, p. 354
30 Bate, *Mysteries of Nature and Art*, p. 82.
31 Ben Jonson, *The Tale of a Tub*, in *The Workes of Benjamin Jonson*, 2 vols (London, 1640), vol. I, p. 76.
32 Boyle Godfrey, *Miscellanea Utilia; or, Miscellaneous Experiments and Observations on Various Subjects*, 2nd edn (London 1737), pp. 68–9.
33 John Mayow, *Medico-physical Works: Being a Translation of Tractatus quinque medico-physici*, trans. A. C. Brown and Leonard Dobbin (Edinburgh, 1907), pp. 75–7.
34 Quoted in Brown, *Big Bang*, p. 12.
35 Buchanan, *Gunpowder, Explosives and the State*, pp. 7–8.
36 Larry Arnold, *Ablaze! The Mysterious Fires of Spontaneous Human Combustion* (New York, 1995).
37 Frederick Marryat, *Jacob Faithful* (Paris, 1837), p. 8.
38 Marryat, *Jacob Faithful*, pp. 7, 6, 8.
39 Emile Zola, *Doctor Pascal*, trans. Ernest A. Vizetelly (Guernsey, 1989), p. 205.
40 Herman Melville, *Redburn: His First Voyage*, ed. Harold Beaver (Harmondsworth, 1976), p. 326.
41 Charles Dickens, *Bleak House*, ed. Norman Page (Harmondsworth, 1972), p. 922.
42 Dickens, *Bleak House*, p. 923.
43 Charles Brockden Brown, *Wieland* and *Memoirs of Carwin the Biloquist*, ed. Jay Fliegelman (London, 1991), p. 18.
44 Brockden Brown, *Wieland* and *Memoirs of Carwin the Biloquist*, p. 19.
45 J. H., *Dreadful news from Limerick being an account of the magazine of powder taking fire the 12th of this instant February, 1694, and the destroying and blowing up of a great part of the city, killing above one hundred* (London, 1694).
46 F. J. Stimson, 'First Harvests', *Scribner's Magazine*, III (1888), p. 160.
47 William Conant Church, 'Nebulae', *The Galaxy*, XXI (February 1876), p. 289.
48 Francis Auguste Koenigstein, *Les Exploits de Ravachol, l'homme à la dynamite* (Paris, 1892–3).
49 Ernest Alfred Vizetelly, *The Anarchists: Their Faith and Their Record, Including Sidelights on the Royal and Other Personages Who Have Been Assassinated* (London, 1911), p. 125.

50 Joseph Conrad, *The Secret Agent: A Simple Tale* (Harmondsworth, 1980), p. 171. References hereafter to *Secret Agent* in text.
51 E. M. Forster, *A Passage to India*, ed. Oliver Stallybrass (London, 1985), p. 144.
52 Walter Benjamin, *Illuminations*, ed. Hannah Arendt, trans. Harry Zohn (London, 1969), p. 236.
53 David Bohm, *Wholeness and the Implicate Order* (London, 1980).
54 'Instantaneous Phenomena Recorded By Photography', *The Manufacturer and Builder*, XI (1879), p. 37.
55 Petr Theodorovich Alisov, *Les Empereurs proposent, la dynamite décompose* (Geneva, 1888).
56 Buchanan, *Gunpowder, Explosives and the State*, p. xxiii.
57 www.evergreen.edu/insideevergreen/911/recollection.htm (accessed 3 March 2004).

ten: The Fizziness Business

1 William Shakespeare, *Hamlet: The Texts of 1603 and 1623*, ed. Ann Thompson and Neil Taylor (London, 2006), III.2, 90, p. 303.
2 John Heydon, 'Epistle Dedicatory', in *Elhavarevna; or, The English Physitians Tutor in the Astrobolismes of Mettals Rosie Cruican* [*sic*] (London, 1665), sig. A5v.
3 Gaston Bachelard, *Air and Dreams: An Essay on the Imagination of Movement*, trans. Edith R. Farrell and C. Frederick Farrell (Dallas, TX, 1988), p. 36.
4 John Milton, *Paradise Lost*, V. 493–9, *The Poems of John Milton*, ed. John Carey and Alistair Fowler (London, 1980), pp. 705–7.
5 Steven Connor, *The Book of Skin* (London, 2003), pp. 257–82.
6 Gaston Bachelard, *La Terre et les rêveries de la volonté* (Paris, 1948), p. 178 [my translation].
7 *The Kalevala: The Epic Poem of Finland*, trans. John Martin Crawford, 2 vols (New York, 1888), vol. I, p. 305.
8 *The Kalevala*, vol. I, p. 311.
9 *The Kalevala*, vol. I, pp. 316–17.
10 Jean Soler, 'The Semiotics of Food in the Bible', in *Food and Drink in History: Selections from the Annales: Economies, Sociétés, Civilisations*, vol. V, ed. Robert Foster and Orest Ranum, trans. Elborg Forster and Patricia M. Ranum (Baltimore and London, 1979), p. 134.
11 W. E. Scudamore, *The Working of the Good Leaven in the People and Church of England*, (London, 1856), p. 3.
12 E. W. Bullinger, *Leaven* (London, 1905), p. 4n.
13 Alfred Austin, *Prince Lucifer* (New York, 1887), p. 118.
14 Bullinger, *Leaven*, p. 13.
15 Alfred Jenour, *The Parable of the Leaven Explained and Applied: Showing Its Important Bearing on the Present Times*, 2nd edn (London, 1855), p. 6.
16 Jenour, *The Parable of the Leaven Explained and Applied*, pp. 5–6.

17 Anon., *Philosophical Enquiry Into Some of the Most Considerable Phenomena's of Nature . . . The Whole Conformable to the Doctrine of Fermentation* (London, 1715), pp. 141–2.
18 Anon., *Philosophical Enquiry*, pp. 152, 153.
19 Basil Valentine, *His Triumphant Chariot of Antinomy With Annotations of Theodore Kirkringius* (1678), ed. L. G. Kelly (New York and London, 1990), p. 25.
20 John Heydon, 'Chymical Dictionary', in *The English Physitians guide; or, A Holy-guide, Leading the Way to Know All Things Past, Present and To Come* (London, 1662), sig. h2v.
21 George Ripley, *The Compound of Alchymy* (London, 1591), sig. I2r.
22 Aristotle, *Generation of Animals: With an English Translation By A. L. Peck* (London and Cambridge, MA, 1943), p. 103.
23 Aristotle, *Generation of Animals.*
24 Aristotle, *Generation of Animals.*
25 Leonard Lessius, *Hygiasticon; or, The Right Course of Preserving Life and Health Unto Extream Old Age Together With Soundnesse and Integritie of the Senses, Judgement, and Memorie*, trans. Nicolas Ferrar (Cambridge, 1634), p. 101.
26 Michael Scot, *The Philosophers Banquet*, 2nd edn (London, 1633), p. 252.
27 Lessius, *Hygiasticon*, pp. 36–7.
28 Jane Huggett, *The Mirror of Health: Food, Diet and Medical Theory, 1450–1660* (Bristol, 1995), pp. 32, 34.
29 Huggett, *The Mirror of Health*, pp. 45, 49.
30 Anon., *The Englishmans Docter; or, The Schoole of Salerne. Or Physicall Observations for the Perfect Preserving of the Body of Man in Continuall Health* , trans. John Harington (London, 1607), sig. A6v.
31 Andrew Boord, *The Breviary of Helthe . . .* (London, 1547), fol. 35r.
32 Huggett, *Mirror of Health*, p. 51.
33 J. C. Drummond and Anne Wilbraham, *The Englishman's Food: A History of Five Centuries of English Diet* (London, 1991), p. 299.
34 T. S. Eliot, *The Complete Poems and Plays* (London, 1969), p. 45.
35 *The Collected Writings of T. E. Hulme*, ed. Karen Csengeri (Oxford, 1994), pp. 62–3; Eliot, *Complete Poems and Plays*, p. 83.
36 Anon., *Directions For Preparing Aerated Medicinal Waters, By Means of the Improved Glass Machines Made at Leith Glass-Works* (Edinburgh, 1787).
37 C. Searle, *Observations on 'Oxygenous Aerated Water', As a Grateful Exhilarating Beverage. And Remedial Agent in Debility, and Depression of the Nervous System; In Asthma and Shortness of Breathing; In Dyspepsia, and Numerous Other Affections. With Some Observations Upon Life: Its Nature and Source* (London, 1839), pp. 7–8.
38 Searle, *Observations on 'Oxygenous Aerated Water'*, p. 9.
39 Immanuel Kant, *The Critique of Judgement*, trans. James Creed Meredith (Oxford, 1957), pp. 199–200.
40 Kant, *The Critique of Judgement*, p. 200.

41 Robin Kingsland, *The Fizziness Business* (London, 1990).
42 Roland Barthes, *Mythologies*, trans. Annette Lavers (London, 1972), p. 37.
43 Barthes, *Mythologies*, p. 37.
44 W. B. Yeats, 'Among School Children', in *Collected Poems* (London and Basingstoke, 1979), p. 244.
45 Barthes, *Mythologies*, pp. 37–8.
46 Margaret Visser, *Much Depends on Dinner: The Extraordinary History and Mythology, Allure and Obsessions, Perils and Taboos of an Ordinary Meal* (New York, 1988), p. 275.
47 Samuel Beckett, *Murphy* (London, 1973), p. 81.
48 Maguelonne Toussaint-Samat, *A History of Food*, trans. Anthea Bell (Oxford, 1992), pp. 284–5.
49 W. B. Yeats, *Collected Plays* (London and Basingstoke, 1982), p. 219.
50 Charles Dickens, *A Christmas Carol, and Other Christmas Books*, ed. Robert Douglas-Fairhurst (Oxford, 2006), p. 21.
51 Chris Waigl, 'Boots-trap: Commentaries', message posted to The Eggcorn Database, (11 August 2005). Online at www.eggcorns.lascribe.net/english/516/boots-trap (accessed 12 April 2010).
52 James Joyce, *Ulysses: The Corrected Text*, ed. Hans Walter Gabler (London, 1986), p. 528.
53 Benjamin Zimmer, 'Figurative Bootstraps', message posted to the American Dialect Society's ADS-L electronic mailing list, 11 August 2005. Online at www.listserv.linguistlist.org/cgi-bin/wa?A2=ind0508b&L=ads-l&P=14972 (accessed 12 April 2010).
54 Fritjof Capra, *The Tao of Physics: An Exploration of the Parallels between Modern Physics and Eastern Mysticism* (London, 1975).
55 Marina Warner, *Phantasmagoria: Spirit Visions, Metaphors and Media into the Twenty-first Century* (Oxford, 2006), p. 381.
56 Warner, *Phantasmagoria*, p. 10.
57 Daniel Tiffany, *Toy Medium: Materialism and Modern Lyric* (Berkeley, CA, and London, 2000). See too Steven Connor, 'Thinking Things', *Textual Practice*, XXIV (2010), pp. 1–20.
58 Jean-Paul Sartre, *Being and Nothingness: An Essay on Phenomenological Ontology*, trans. Hazel E. Barnes (London, 1984), p. 329.
59 Steven Connor, 'Next to Nothing', *Tate Etc*, 12 (2008), pp. 82–93.

Further Reading

Anon., *Philosophical Enquiry into Some of the Most Considerable Phenomena's of Nature . . . The Whole Conformable to the Doctrine of Fermentation* (London, 1715)

Anon., *An Heroic Epistle to Mr Winsor, The Patentee of the Hydro-Carbonic Gas Lights and Founder of the National Light and Heat Company* (London, 1808)

Aristotle, *Meteorologica*, trans. E. W. Webster, in *The Works of Aristotle Translated into English*, ed. W. D. Ross, vol. III (Oxford, 1923)

Bachelard, Gaston, *Air and Dreams: An Essay on the Imagination of Movement*, trans. Edith R. Farrell and C. Frederick Farrell (Dallas, TX, 1988)

Barrow, John D., *The Book of Nothing* (London, 2001)

Beddoes, Thomas, *Notice of Some Observations Made at the Medical Pneumatic Institution* (Bristol, 1793)

Boyle, Robert, *The Works of Robert Boyle*, ed. Michael Hunter and Edward B. Davis, 14 vols (London, 1999–2000)

Brimblecombe, Peter, *The Big Smoke: A History of Air Pollution in London since Medieval Times* (London, 1987)

Brown, G. I., *The Big Bang: A History of Explosives* (Stroud, 2005)

Bullinger, E. W., *Leaven* (London, 1905)

Cantor, G. N., and M.J.S. Hodge, eds, *Conceptions of Ether: Studies in the History of Ether Theories, 1740–1900* (Cambridge, 1981)

Catlin, George, *The Breath of Life; or, Mal-Respiration and its Effects Upon the Enjoyments and Life of Man* (London, 1861)

Chandler, Dean, *Outline of History of Lighting by Gas* (London, 1936)

Connor, Steven, 'Building Breathing Space', in *Going Aerial: Air, Art, Architecture*, ed. Monika Bakke (Maastricht, 2007), pp. 118–34

——, 'Next to Nothing.' *Tate Etc*, XII (2008), pp. 82–93

——, 'Atmospheres: Imagining Air', http://www.stevenconnor.com/atmospheres.htm

Crosland, Maurice, '"Slippery Substances": Some Practical and Conceptual Problems in the Understanding of Gases in the Pre-Lavoisier Era', in *Instruments and Experimentation in the History of Chemistry*, ed. Frederic L. Holmes and Trevor H. Levere (Cambridge, MA, and London, 2000), pp. 78–104

Davy, Humphrey, *Researches, Chemical and Philosophical; Chiefly Concerning Nitrous Oxide; or, Dephlogisticated Nitrous Air, and its Respiration* (London, 1800)
Gilman, Sander L., and Zhou Xun, eds, *Smoke: A Global History of Smoking* (London, 2004)
Glinsky, Albert, *Theremin: Ether Music and Espionage* (Urbana, and Chicago, IL, 2000)
Haber, L. F., *The Poisonous Cloud: Chemical Warfare in the First World War* (Oxford, 1986)
Hales, Stephen, *Vegetable Staticks; or, An Account of Some Statical Experiments on the Sap in Vegetables: Being an Essay Towards a Natural History of Vegetation. Also, a Specimen of an Attempt to Analyse the Air, by a Great Variety of Chymio-statical Experiments* (London, 1727)
Haslam, John, *Illustrations of Madness*, ed. Roy Porter (London and New York, 1988)
Hero of Alexandria, *The Pneumatics*, trans. Bennet Woodcroft (London, 1851)
Irigaray, Luce, *The Forgetting of Air in Martin Heidegger*, trans. Mary Beth Mader (London, 1999).
James, William, 'On Some Hegelisms', *Mind*, VII (1982), pp. 186–208
Jay, Mike, *The Air-Loom Gang; The Strange and True Story of James Tilly Matthews and his Visionary Madness* (London and New York, 2004)
——, *The Atmosphere of Heaven: The Unnatural Experiments of Dr Beddoes and his Sons of Genius* (New Haven, CT, and London, 2009)
Johnson, Stephen, *The Invention of Air: An Experiment, a Journey, a New Country and the Amazing Force of Scientific Discovery* (London, 2009)
Krauß, Friedrich, *Nothschrei eines Magnetisch-Vergifteten (1852) und Nothgedrungene Fortsetzung meines Nothschrei (1867): Selbstschilderungen eines Geisteskranken*, ed. H. Ahlenstiel and J. E. Meyer (Göttingen, 1967)
Mayow, John. *Medico-physical Works: Being a Translation of Tractatus quinque medico-physici*, trans. A. C. Brown and Leonard Dobbin (Edinburgh, 1907)
Mesmer, F. A., *Mesmerism: A Translation of the Original Scientific and Medical Writings of F. A. Mesmer*, trans. George Bloch (Los Altos, CA, 1980)
Newton, David E., *Encyclopedia of Air* (Westport, CT, 2003)
Perceval, John, *Perceval's Narrative: A Patient's Account of his Psychosis, 1830–1832*, ed. Gregory Bateson (Stanford, CA, 1961)
Prothero, Stephen, *Purified By Fire: A History of Cremation in America* (Berkeley, Los Angeles and London, 2001)
Polli, Andrea, '"Atmospherics/Weather Works": A Spatialized Meteorological Data Sonification Project', *Leonardo*, XXXVIII (2005), pp. 31–6
Polwhele, Richard, 'The Pneumatic Revellers: An Eclogue', *Poems*, 5 vols (Truro, 1810), vol. V, pp. iii–viii, 1–17
Priestley, Joseph, *Experiments and Observations on Different Kinds of Air*, 3 vols (London, 1774–7)
Raudive, Konstantin, *Breakthrough: An Amazing Experiment in Electronic Communication with the Dead*, trans. Nadia Fowler, ed. Joyce Morton (Gerrards Cross, 1971)

Richardson, Benjamin Ward, 'The Theory of a Nervous Ether', in *Half-Hour Recreations in Popular Science*, 1st Series, ed. Diana Estes (Boston, MA, 1874), pp. 362–74

Robinson, Thomas, *Breathing: Considered in Relation to the Bodily, Mental, and Social Life of Man* (Glasgow, 1869)

Sala, George Augustus, 'The Secrets of the Gas', in *Gaslight and Daylight: Some London Scenes They Shine Upon* (London, 1859), pp. 156–63

Schivelbusch, Wolfgang, *Disenchanted Night: The Industrialisation of Light in the Nineteenth Century*, trans. Angela Davies (Oxford, New York and Hamburg, 1988)

Sloterdijk, Peter, *Schäume: Sphären, vol. 3: Plurale Sphärologie* (Frankfurt, 2004)

——, *Terror from the Air*, trans. Amy Patton and Steve Corcoran (Cambridge, MA, and London, 2009)

Smith, W.D.A., *Under the Influence: A History of Nitrous Oxide and Oxygen Anaesthesia* (London and Basingstoke, 1982)

Van Helmont, Jean-Baptiste, *Van Helmont's Works*, trans. J.C. (London, 1664)

Walker, Gabrielle, *An Ocean of Air: A Natural History of the Atmosphere* (London, 2008)

Warner, Marina, *Phantasmagoria: Spirit Visions, Metaphors and Media into the Twenty-first Century* (Oxford, 2006)

Wells, H. G., *The War in the Air: and Particularly How Mr Bert Smallways Fared While it Lasted* (London, 2002)

Photo Acknowledgements

The author and publishers wish to express their thanks to the below sources of illustrative material and/or permission to reproduce it:

From Robert Boyle, *A Continuation of New Experiments Physico-Mechanical, Touching the Spring and Weight of the Air and their Effects* (Oxford, 1669): p. 19; from John Clubbe, *Physiognomy: Being a Sketch Only of a Larger Work Upon the Same Plan: Wherein the Different Tempers, Passions, and Manners of Men, Will Be Particularly Considered* (London, 1763): p. 337; from *Dr Syntax in Paris; or a Tour in the Search of the Grotesque; being a Humorous Delineation of the Pleasures and Miseries of the French Metropolis* (London, 1820): p. 111; from Gustave Doré and Blanchard Jerrold, *London: A Pilgrimage* (London, 1872): p. 137; Fogg Art Museum, Harvard University, Cambridge, Mass.: p. 188; from John Joseph Griffin, *Chemical Recreations: A Series of Amusing and Instructive Experiments, which may be Performed with Ease, Safety, Success, and Economy . . .*, 7th edn (Glasgow and London, 1834): p. 69; from James Haslam, *Illustrations of Madness: Exhibiting a Singular Case of Insanity, And a No Less Remarkable Difference in Medical Opinion: Developing the Nature of An Assailment, And the Manner of Working Events; with a Description of Tortures Experienced by Bomb-Bursting, Lobster-Cracking and Lengthening the Brain* (London, 1810): p. 49; from Jean Ingen-Housz, *Nouvelles expériences et observations sur divers objets de physique* vol. I (Paris, 1785): p. 67; from William Kirkby, *The Evolution of Artificial Mineral Waters* (Manchester, 1902): p. 326; from Francis Malthus, *A Treatise of Artificial Fire-works Both for Warres and Recreation: with divers pleasant Geometricall observations, Fortifications, and Arithmeticall examples* (London, 1629): p. 289; from Benjamin Martin, *The Description and Use of a New, Portable, Table Air-Pump and Condensing Engine. With a Select Variety of Capital Experiments . . .* (London, 1766): p.24; Musée des Beaux-Arts, Lyon: p. 187; National Gallery of Canada, Ottawa: p. 185; from Joseph Priestley, *Experiments and Observations on Different Kinds of Air . . . The Second Edition Corrected* (London, 1775): p. 64; from John Scoffern, *Chemistry No Mystery; or, a Lecturer's Bequest* (London, 1839): p. 163; Tate, London: pp. 184, 186; from Robert Ward, *Animadversions of Warre, or, A Militarie Magazine of the Truest Rules, and Ablest Instructions, for the Managing of Warre* (London, 1639): p. 295.

Index